Introduction to

FEEDBACK

CONTROL

THEORY

Introduction to

FEEDBACK

CONTROL

THEORY

Hitay Özbay

The Ohio State University

Columbus, Ohio

CRC Press

Taylor & Francis Group

Boca Raton London New York

CRC Press is an imprint of the
Taylor & Francis Group, an **informa** business

CRC Press
Taylor & Francis Group
6000 Broken Sound Parkway NW, Suite 300
Boca Raton, FL 33487-2742

First issued in paperback 2019

ISBN-13: 978-0-8493-1867-2 (hbk)
ISBN-13: 978-0-367-39953-5 (pbk)

Library of Congress Card Number 99–33365

Library of Congress Cataloging-in-Publication Data

Özbay, Hitay
 Introduction to feedback control theory / Hitay Özbay.
 p. cm.
 Includes bibliographical references.
 ISBN 0-8493-1867–X (alk. paper)
 1. Feedback control systems. I.Title.
TJ216.O97 1999
629.8′ 32--dc21

 99–33365
 CIP

Visit the Taylor & Francis Web site at
http://www.taylorandfrancis.com

and the CRC Press Web site at
http://www.crcpress.com

Preface

This book is based on my lecture notes for a ten-week second course on feedback control systems. In our department the first control course is at the junior level; it covers the basic concepts such as dynamical systems modeling, transfer functions, state space representations, block diagram manipulations, stability, Routh-Hurwitz test, root locus, lead-lag controllers, and pole placement via state feedback. In the second course, (open to graduate and undergraduate students) we review these topics briefly and introduce the Nyquist stability test, basic loopshaping, stability robustness (Kharitanov's theorem and its extensions, as well as \mathcal{H}^{∞}-based results) sensitivity minimization, time delay systems, and parameterization of all stabilizing controllers for single input-single output (SISO) stable plants. There are several textbooks containing most of these topics, e.g. [7, 17, 22, 37, 45]. But apparently there are not many books covering *all* of the above mentioned topics. A slightly more advanced text that I would especially like to mention is *Feedback Control Theory*, by Doyle, Francis, and Tannenbaum, [18]. It is an excellent book on SISO \mathcal{H}^{∞}-based robust control, but it is lacking significant portions of the *introductory* material included in our curriculum. I hope that the present book fills this gap, which may exist in other universities as well.

It is also possible to use this book to teach a course on feedback control, following a one-semester signals and systems course based on [28, 38], or similar books dedicating a couple of chapters to control-related topics. To teach a one-semester course from the book, Chapter 11 should be expanded with supplementary notes so that the state space methods are covered more rigorously.

Now a few words for the students. The exercise problems at the end of each chapter may or may not be similar to the examples given in the text. You should first try solving them by hand calculations; if you think that a computer-based solution is the only way, then go ahead and use MATLAB. I assume that you are familiar with MATLAB; for those who are not, there are many introductory books, e.g., [19, 23, 44]. Although it is not directly related to the present book, I would also recommend [52] as a good reference on MATLAB-based computing.

Despite our best efforts, there may be errors in the book. Please send your comments to: ozbay.1@osu.edu, I will post the corrections on the web: http://eewww.eng.ohio-state.edu/~ozbay/ifct.html.

Many people have contributed to the book directly or indirectly. I would like to acknowledge the encouragement I received from my colleagues in the Department of Electrical Engineering at The Ohio State University, in particular J. Cruz, H. Hemami, Ü. Özgüner, K. Passino, L. Potter, V. Utkin, S. Yurkovich, and Y. Zheng. Special thanks to A. Tannenbaum for his encouraging words about the potential value of this book. Students who have taken my courses have helped significantly with their questions and comments. Among them, R. Bhojani and R. Thomas read parts of the latest manuscript and provided feedback. My former PhD students T. Peery, O. Toker, and M. Zeren helped my research; without them I would not have been able to allocate extra time to prepare the supplementary class notes that eventually formed the basis of this book. I would also like to acknowledge National Science Foundation's support of my current research. The most significant direct contribution to this book came from my wife Özlem, who was always right next to me while I was writing. She read and criticized the preliminary versions of the book. She also helped me with the MATLAB plots. Without her support, I could not have found the motivation to complete this project.

<div align="right">
Hitay Özbay

Columbus, May 1999
</div>

Dedication

To my wife, Özlem

Contents

Chapter 1

Introduction

1.1 Feedback Control Systems

Examples of feedback are found in many disciplines such as engineering, biological sciences, business, and economy. In a feedback system there is a process (a cause-effect relation) whose operation depends on one or more variables (inputs) that cause changes in some other variables. If an input variable can be manipulated, it is said to be a control input, otherwise it is considered a disturbance (or noise) input. Some of the process variables are monitored; these are the outputs. The feedback controller gathers information about the process behavior by observing the outputs, and then it generates the new control inputs in trying to make the system behave as desired. Decisions taken by the controller are crucial; in some situations they may lead to a catastrophe instead of an improvement in the system behavior. This is the main reason that feedback controller design (i.e., determining the rules for automatic decisions taken by the feedback controller) is an important topic.

A typical feedback control system consists of four subsystems: a *process* to be controlled, sets of *sensors* and *actuators*, and a *controller*,

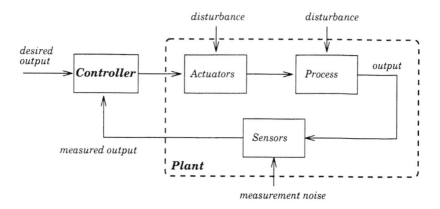

Figure 1.1: Feedback control system.

as shown in Figure 1.1. The process is the actual physical system that cannot be modified. Actuators and sensors are selected by process engineers based on physical and economical constraints (i.e., the range of signals to be measured and/or generated and accuracy versus cost of these devices). The controller is to be designed for a given *plant* (the overall system, which includes the process, sensors, and actuators).

In engineering applications the controller is usually a computer, or a human operator interfacing with a computer. Biological systems can be more complex; for example, the central nervous system is a very complicated controller for the human body. Feedback control systems encountered in business and economy may involve teams of humans as main decision makers, e.g., managers, bureaucrats, and/or politicians.

A good understanding of the process behavior (i.e., the cause-effect relationship between input and output variables) is extremely helpful in designing the rules for control actions to be taken. Many engineering systems are described accurately by the physical laws of nature. So, mathematical models used in engineering applications contain relatively low levels of uncertainty, compared with mathematical models that appear in other disciplines, where input-output relationships

can be much more complicated.

In this book, certain fundamental problems of feedback control theory are studied. Typical application areas in mind are in engineering. It is assumed that there is a mathematical model describing the dynamical behavior of the underlying process (modeling uncertainties will also be taken into account). Most of the discussion is restricted to single input-single output (SISO) processes. An important point to keep in mind is that success of the feedback control depends heavily on the accuracy of the process/uncertainty model, whether this model captures the reality or not. Therefore, the first step in control is to derive a simple and relatively accurate mathematical model of the underlying process. For this purpose, control engineers must communicate with process engineers who know the physics of the system to be controlled. Once a mathematical model is obtained and performance objectives are specified, control engineers use certain design techniques to synthesize a feedback controller. Of course, this controller must be tested by simulations and experiments to verify that performance objectives are met. If the achieved performance is not satisfactory, then the process model and the design goals must be reevaluated and a new controller should be designed from the new model and the new performance objectives. This iteration should continue until satisfactory results are obtained, see Figure 1.2.

Modeling is a crucial step in the controller design iterations. The result of this step is a nominal process model and an uncertainty description that represents our confidence level for the nominal model. Usually, the uncertainty magnitude can be decreased, i.e., the confidence level can be increased only by making the nominal plant model description more complicated (e.g., increasing the number of variables and equations). On the other hand, controller design and analysis for very complicated process models are very difficult. This is the basic trade-off in system modeling. A useful nominal process model should

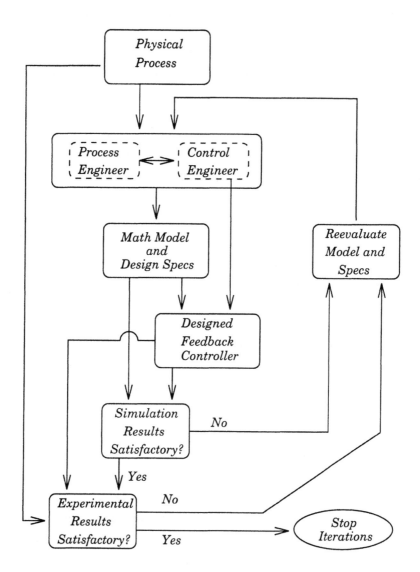

Figure 1.2: Controller design iterations.

Figure 1.3: A MIMO system.

be simple enough so that the controller design is feasible. At the same time the associated uncertainty level should be low enough to allow the performance analysis (simulations and experiments) to yield acceptable results.

The purpose of this book is to present basic feedback controller design and analysis (performance evaluation) techniques for simple SISO process models and associated uncertainty descriptions. Examples from certain specific engineering applications will be given whenever it is necessary. Otherwise, we will just consider generic mathematical models that appear in many different application areas.

1.2 Mathematical Models

A multi-input-multi-output (MIMO) system can be represented as shown in Figure 1.3, where u_1, \ldots, u_p are the inputs and y_1, \ldots, y_q are the outputs (for SISO systems we have $p = q = 1$). In this figure, the direction of the arrows indicates that the inputs are processed by the system to generate the outputs.

In general, feedback control theory deals with dynamical systems, i.e., systems with internal memory (in the sense that the output at time $t = t_0$ depends on the inputs applied at time instants $t \leq t_0$). So, the plant models are usually in the form of a set of differential equations obtained from physical laws of nature. Depending on the operating conditions, input/output relation can be best described by linear or

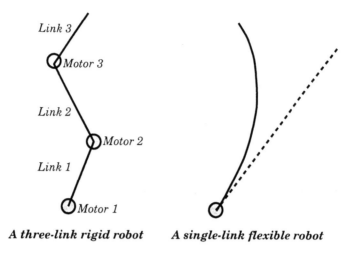

A three-link rigid robot **A single-link flexible robot**

Figure 1.4: Rigid and flexible robots.

nonlinear, partial or ordinary differential equations.

For example, consider a three-link robot as shown in Figure 1.4. This system can also be seen as a simple model of the human body. Three motors located at the joints generate torques that move the three links. Position, and/or velocity, and/or acceleration of each link can be measured by sensors (e.g., optical light with a camera, or gyroscope). Then, this information can be processed by a feedback controller to produce the rotor currents that generate the torques. The feedback loop is hence closed. For a successful controller design, we need to understand (i.e., derive mathematical equations of) how torques affect position and velocity of each link, and how current inputs to motors generate torque, as well as the sensor behavior. The relationship between torque and position/velocity can be determined by laws of physics (Newton's law). If the links are rigid, then a set of nonlinear ordinary differential equations is obtained, see [26] for a mathematical model. If the analysis and design are restricted to small displacements around the upright equilibrium, then equations can be linearized without introducing too much error [29]. If the links are made of a flexible material (for example,

in space applications the weight of the material must be minimized to reduce the payload, which forces the use of lightweight flexible materials), then we must consider bending effects of the links, see Figure 1.4. In this case, there are an infinite number of position coordinates, and partial differential equations best describe the overall system behavior [30, 46].

The robotic examples given here show that a mathematical model can be *linear* or *nonlinear*, *finite dimensional* (as in the rigid robot case) or *infinite dimensional* (as in the flexible robot case). If the parameters of the system (e.g., mass and length of the links, motor coefficients, etc.) do not change with time, then these models are *time-invariant*, otherwise they are *time-varying*.

In this book, *linear time-invariant* (LTI) models will be considered only. Most of the discussion will be restricted to *finite dimensional* models, but certain simple infinite dimensional models (in particular time delay systems) will also be discussed.

The book is organized as follows. In Chapter 2, modeling issues and sources of uncertainty are studied and the main reason to use feedback is explained. Typical performance objectives are defined in Chapter 3. In Chapter 4, basic stability tests are given. Single parameter controller design is covered in Chapter 5 by using the root locus technique. Stability robustness and stability margins are defined in Chapter 6 via Nyquist plots. Stability analysis for systems with time delays is in Chapter 7. Simple lead-lag and PID controller design methods are discussed in Chapter 8. Loopshaping ideas are introduced in Chapter 9. In Chapter 10, robust stability and performance conditions are defined and an \mathcal{H}^∞ controller design procedure is outlined. Finally, state space based controller design methods are briefly discussed and a parameterization of all stabilizing controllers is presented in Chapter 11.

Chapter 2

Modeling, Uncertainty, and Feedback

2.1 Finite Dimensional LTI System Models

Throughout the book, linear time-invariant (LTI) single input–single output (SISO) plant models are considered. Finite dimensional LTI models can be represented in time domain by dynamical state equations in the form

$$\dot{x}(t) = Ax(t) + Bu(t) \tag{2.1}$$
$$y(t) = Cx(t) + Du(t) \tag{2.2}$$

where $y(t)$ is the output, $u(t)$ is the input, and the components of the vector $x(t)$ are state variables. The matrices A, B, C, D form a state space realization of the plant. Transfer function $P(s)$ of the plant is the frequency domain representation of the input-output behavior:

$$Y(s) = P(s)\, U(s)$$

where s is the Laplace transform variable, $Y(s)$ and $U(s)$ represent the Laplace transforms of $y(t)$ and $u(t)$, respectively. The relation between state space realization and the transfer function is

$$P(s) = C(sI - A)^{-1}B + D.$$

Transfer function of an LTI system is unique, but state space realizations are not.

Consider a generic finite dimensional SISO transfer function

$$P(s) = K_p \frac{(s - z_1)\ldots(s - z_m)}{(s - p_1)\ldots(s - p_n)} \tag{2.3}$$

where z_1, \ldots, z_m are the zeros and p_1, \ldots, p_n are the poles of $P(s)$. Note that for causal systems $n \geq m$ (i.e., direct derivative action is not allowed) and in this case, $P(s)$ is said to be a *proper* transfer function since it satisfies

$$|d| := \lim_{|s| \to \infty} |P(s)| < \infty. \tag{2.4}$$

If $|d| = 0$ in (2.4) then $P(s)$ is *strictly proper*. For proper transfer functions, (2.3) can be rewritten as

$$P(s) = \frac{b_1 s^{n-1} + \ldots + b_n}{s^n + a_1 s^{n-1} + \ldots + a_n} + d. \tag{2.5}$$

The state space realization of (2.5) in the form

$$A = \begin{bmatrix} 0_{(n-1)\times 1} & I_{(n-1)\times(n-1)} \\ -a_n & \cdots & -a_1 \end{bmatrix} \quad B = \begin{bmatrix} 0_{(n-1)\times 1} \\ 1 \end{bmatrix}$$
$$C = \begin{bmatrix} b_n & \cdots & b_1 \end{bmatrix} \quad D = d.$$

is called the controllable canonical realization. In this book, transfer function representations will be used mostly. A brief discussion on state space based controller design is included in the last chapter.

2.2 Infinite Dimensional LTI System Models

Multidimensional systems, spatially distributed parameter systems and systems with time delays are typical examples of infinite dimensional systems. Transfer functions of infinite dimensional LTI systems can not be written as rational functions in the form (2.5). They are either transcendental functions, or infinite sums, or products, of rational functions. Several examples are given below; for additional examples see [12]. For such systems state space realizations (2.1, 2.2) involve infinite dimensional operators A, B, C, (i.e., these are not finite-size matrices) and an infinite dimensional state $x(t)$ which is not a finite-size vector. See [13] for a detailed treatment of infinite dimensional linear systems.

2.2.1 A Flexible Beam

A flexible beam with free ends is shown in Figure 2.1. The deflection at a point x along the beam and at time instant t is denoted by $w(x,t)$. Ideal Euler-Bernoulli beam model with Kelvin-Voigt damping, [34, 46], is a reasonably simple mathematical model:

$$\rho \frac{\partial^2 w}{\partial t^2} + 2\alpha \frac{\partial^2}{\partial x^2} \left(EI \frac{\partial^3 w}{\partial x^2 \partial t} \right) + \frac{\partial^2}{\partial x^2} \left(EI \frac{\partial^2 w}{\partial x^2} \right) = 0, \qquad (2.6)$$

where $\rho(x)$ denotes the mass density per unit length of the beam, $EI(x)$ denotes the second moment of the modulus of elasticity about the elastic axis and $\alpha > 0$ is the damping factor.

Let $2\alpha = \epsilon > 0$, $\rho = 1$, $EI = 1$, and suppose that a transverse force, $-u(t)$, is applied at one end of the beam, $x = 1$, and the deflection at the other end is measured, $y(t) = w(0,t)$. Then, the boundary conditions for (2.6) are

$$\frac{\partial^2 w}{\partial x^2}(0,t) + \epsilon \frac{\partial^3 w}{\partial x^2 \partial t}(0,t) = 0, \quad \frac{\partial^2 w}{\partial x^2}(1,t) + \epsilon \frac{\partial^3 w}{\partial x^2 \partial t}(1,t) = 0,$$

Figure 2.1: A flexible beam with free ends.

$$\frac{\partial^3 w}{\partial x^3}(0,t) + \epsilon \frac{\partial^4 w}{\partial x^3 \partial t}(0,t) = 0, \quad \frac{\partial^3 w}{\partial x^3}(1,t) + \epsilon \frac{\partial^4 w}{\partial x^3 \partial t}(1,t) = u(t).$$

By taking the Laplace transform of (2.6) and solving the resulting fourth-order ODE in x, the transfer function $P(s) = Y(s)/U(s)$ is obtained (see [34] for further details):

$$P(s) = \frac{1}{(1+\epsilon s)\beta^3}\left(\frac{\sinh \beta - \sin \beta}{\cos \beta \cosh \beta - 1}\right)$$

where $\beta^4 = \frac{-s^2}{(1+\epsilon s)}$. It can be shown that

$$P(s) = \frac{2}{s^2}\prod_{n=1}^{\infty}\left(\frac{1+\epsilon s - \frac{s^2}{4\alpha_n^4}}{1+\epsilon s + \frac{s^2}{\phi_n^4}}\right). \tag{2.7}$$

The coefficients α_n, ϕ_n, $n = 1, 2, \ldots$, are the roots of

$$\cos \alpha_n \sinh \alpha_n = \sin \alpha_n \cosh \alpha_n \quad \text{and} \quad \cos \phi_n \cosh \phi_n = 1$$

for $\alpha_n, \phi_n > 0$. Let $\phi_j < \phi_k$ and $\alpha_j < \alpha_k$ for $j < k$, then ϕ_n's alternate with α_n's. It is also easy to show that $\phi_n \to \frac{\pi}{2} + n\pi$ and $\alpha_n \to \frac{\pi}{4} + n\pi$ as $n \to \infty$.

2.2.2 Systems with Time Delays

In some applications, information flow requires significant amounts of time delay due to physical distance between the process and the controller. For example, if a spacecraft is controlled from the earth, meas-

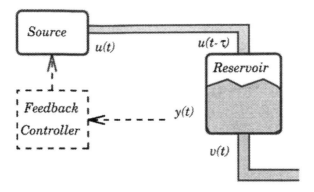

Figure 2.2: Flow control problem.

urements and command signals reach their destinations with a non-negligible time delay even though signals travel at (or near) the speed of light. There may also be time delays within the process, or the controller itself (e.g., when the controller is very complicated, computations may take a relatively long time introducing computational time delays).

As an example of a time delay system, consider a generic flow control problem depicted in Figure 2.2, where $u(t)$ is the input flow rate at the source, $v(t)$ is outgoing flow rate, and $y(t)$ is the accumulation at the reservoir. This setting is also similar to typical data flow control problems in high speed communication networks, where packet flow rates at sources are controlled to keep the queue size at bottleneck node at a desired level, [41]. A simple mathematical model is:

$$\dot{y}(t) = u(t - \tau) - v(t)$$

where τ is the travel time from source to reservoir. Note that, to solve this differential equation for $t > 0$, we need to know $y(0)$ and $u(t)$ for $t \in (-\tau , 0]$, so an infinite amount of information is required. Hence, the system is infinite dimensional. In this example, $u(t)$ is adjusted by the feedback controller; $v(t)$ can be known or unknown to the controller, in the latter case it is considered a disturbance. The output

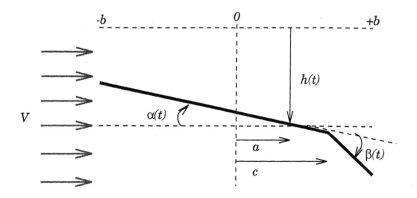

Figure 2.3: A thin airfoil.

is $y(t)$. Assuming zero initial conditions, the system is represented in the frequency domain by

$$Y(s) = \frac{1}{s} \left(e^{-\tau s} U(s) - V(s) \right).$$

The transfer function from u to y is $\frac{1}{s} e^{-\tau s}$. Note that it contains the time delay term $e^{-\tau s}$, which makes the system infinite dimensional.

2.2.3 Mathematical Model of a Thin Airfoil

Aeroelastic behavior of a thin airfoil, shown in Figure 2.3, is also represented as an infinite dimensional system. If the air-flow velocity, V, is higher than a certain critical speed, then unbounded oscillations occur in the structure; this is called the flutter phenomenon. Flutter suppression (stabilizing the structure) and gust alleviation (reducing the effects of sudden changes in V) problems associated with this system are solved by using feedback control techniques, [40, 42]. Let $z(t) := [h(t), \alpha(t), \beta(t)]^{\mathsf{T}}$ and $u(t)$ denote the control input (torque applied at the flap).

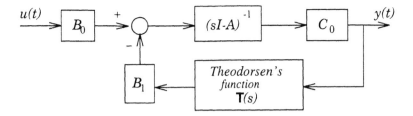

Figure 2.4: Mathematical model of a thin airfoil.

For a particular output in the form

$$y(t) = c_1 z(t) + c_2 \dot{z}(t)$$

the transfer function is

$$\frac{Y(s)}{U(s)} = P(s) = \frac{C_0(sI - A)^{-1} B_0}{1 - C_0(sI - A)^{-1} B_1 \ \mathsf{T}(s)}$$

where $C_0 = [c_1 \ c_2]$, and A, B_0, B_1 are constant matrices of appropriate dimensions (they depend on V, a, b, c, and other system parameters related to the geometry and physical properties of the structure) and $\mathsf{T}(s)$ is the so-called Theodorsen's function, which is a minimum phase stable causal transfer function

$$\mathrm{Re}(\mathsf{T}(j\omega)) = \frac{J_1(\omega_r)(J_1(\omega_r) + Y_0(\omega_r)) + Y_1(\omega_r)(Y_1(\omega_r) - J_0(\omega_r))}{(J_1(\omega_r) + Y_0(\omega_r))^2 + (Y_1(\omega_r) - J_0(\omega_r))^2}$$

$$\mathrm{Im}(\mathsf{T}(j\omega)) = \frac{-(Y_1(\omega_r)Y_0(\omega_r) + J_1(\omega_r)J_0(\omega_r))}{(J_1(\omega_r) + Y_0(\omega_r))^2 + (Y_1(\omega_r) - J_0(\omega_r))^2}$$

where $\omega_r = \omega \, b/V$ and J_0, J_1, Y_0, Y_1 are Bessel functions. Note that the plant itself is a feedback system with infinite dimensional term $\mathsf{T}(s)$ appearing in the feedback path, see Figure 2.4.

2.3 Linearization of Nonlinear Models

2.3.1 Linearization Around an Operating Point

Linear models are sometimes obtained by linearizing nonlinear differential equations around an operating point. To illustrate the linearization procedure, consider a generic nonlinear term in the form

$$\dot{x}(t) = f(x(t))$$

where $f(\cdot)$ is an analytic function around a point x_e. Suppose that x_e is an equilibrium point: i.e., $f(x_e) = 0$, so that if $x(t_0) = x_e$ then $x(t) = x_e$ for all $t \geq t_0$. Let δ_x represent small deviations from x_e and consider the system behavior at $x(t) = x_e + \delta_x(t)$:

$$\dot{x}(t) = \dot{\delta}_x(t) = f(x_e + \delta_x(t)) = f(x_e) + \left(\frac{\partial f}{\partial x}\right)_{x=x_e} \delta_x(t) + H.O.T.$$

where $H.O.T.$ represents the higher-order terms involving $(\delta_x)^2/2!$, $(\delta_x)^3/3!$, ..., and higher-order derivatives of f with respect to x evaluated at $x = x_e$. As $|\delta_x| \to 0$ the effect of higher-order terms are negligible and the dynamical equation is approximated by

$$\dot{\delta}_x(t) = A\delta_x(t) \qquad \text{where} \qquad A := \left(\frac{\partial f}{\partial x}\right)_{x=x_e}$$

which is a linear system.

Example 2.1 The equations of motion of the pendulum shown in Figure 2.5 are given by Newton's law: *mass times acceleration is equal to the total force*. The gravitational-force component along the direction of the rod is canceled by the reaction force. So the pendulum swings in the direction orthogonal to the rod: In this coordinate, the acceleration is $\ell\ddot{\theta}$ and the gravitational force is $-mg\sin(\theta)$ (it is in the opposite

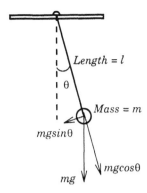

Figure 2.5: A free pendulum.

direction to θ). Assuming there is no friction, equations of motion are

$$
\begin{aligned}
\dot{x}_1(t) &= x_2(t) \\
\dot{x}_2(t) &= -\frac{mg}{\ell}\sin(x_1(t))
\end{aligned}
$$

where $x_1(t) = \theta(t)$ and $x_2(t) = \dot{\theta}(t)$, and $x(t) = [x_1(t) \quad x_2(t)]^\mathsf{T}$ is the state vector. Clearly $x_e = [0 \quad 0]^\mathsf{T}$ is an equilibrium point. When $|\theta(t)|$ is small, the nonlinear term $\sin(x_1(t))$ is approximated by $x_1(t)$. So the linearized equations lead to

$$
\ddot{x}_1(t) = -\frac{mg}{\ell}x_1(t).
$$

For an initial condition $x(0) = [\theta_o \quad 0]^\mathsf{T}$, (where $0 < |\theta_o| \approx 0$), the pendulum oscillates sinusoidally with natural frequency $\sqrt{\frac{mg}{\ell}}\,\mathrm{rad/sec}$.

2.3.2 Feedback Linearization

Another way to obtain a linear model from a nonlinear system equation is *feedback linearization*. The basic idea of feedback linearization is illustrated in Figure 2.6. The nonlinear system, whose state is x, is linearized by using a nonlinear feedback to generate an appropriate input u. The closed-loop system from external input r to output x is

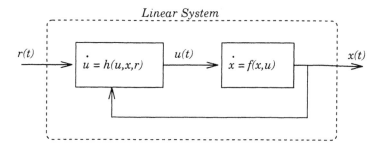

Figure 2.6: Feedback linearization.

linear. Feedback linearization rely on precise cancelations of certain nonlinear terms, therefore it is not a robust scheme. Also, for certain types of nonlinear systems, a linearizing feedback does not exist. See e.g. [27, 31] for analysis and design of nonlinear feedback systems.

Example 2.2 Consider the inverted pendulum system shown in Figure 2.7. This is a classical feedback control example; it appears in almost every control textbook. See for example [37, pp. 85–87] where typical system parameters are taken as

$$m = 0.1 \text{ kg}, \quad M = 2 \text{ kg}, \quad \ell = 0.5 \text{ m}, \quad g = 9.8 \text{ m/sec}^2, \quad J = m\ell^2/3.$$

The aim here is to balance the stick at the upright equilibrium point, $\theta = 0$, by applying a force $u(t)$ that uses feedback from $\theta(t)$ and $\dot{\theta}(t)$. The equations of motion can be written from Newton's law:

$$(J + \ell^2 m)\ddot{\theta} + \ell m \cos(\theta)\ddot{x} - \ell m g \sin(\theta) \quad = \quad 0 \qquad (2.8)$$

$$(M + m)\ddot{x} + m\ell \cos(\theta)\ddot{\theta} - m\ell \sin(\theta)\dot{\theta}^2 \quad = \quad u \; . \qquad (2.9)$$

By using equation (2.9), \ddot{x} can be eliminated from equation (2.8) and hence a direct relationship between θ and u can be obtained as

$$\left(\frac{J}{m\ell} + \ell - \frac{m\ell \cos^2(\theta)}{M + m} \right) \ddot{\theta} + \frac{m\ell \cos(\theta) \sin(\theta)\dot{\theta}^2}{M + m} - g \sin(\theta)$$

$$= -\frac{\cos(\theta)}{M + m} u. \qquad (2.10)$$

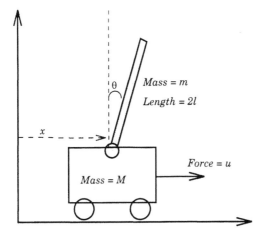

Figure 2.7: Inverted pendulum on a cart.

Note that if $u(t)$ is chosen as the following nonlinear function of $\theta(t)$ and $\dot{\theta}(t)$

$$u = -\frac{M+m}{\cos(\theta)}\left(\frac{m\ell\cos(\theta)\sin(\theta)\dot{\theta}^2}{M+m} - g\sin(\theta)\right.$$

$$\left. -(\frac{J}{m\ell}+\ell-\frac{m\ell\cos^2(\theta)}{M+m})\left(\alpha\dot{\theta}+\beta\theta-r_\theta\right)\right) \quad (2.11)$$

then θ satisfies the equation

$$\ddot{\theta} + \alpha\dot{\theta} + \beta\theta = r_\theta \quad\quad\quad (2.12)$$

where α and β are the parameters of the nonlinear controller and $r_\theta(t)$ is the reference input, i.e., desired $\theta(t)$. The equation (2.12) represents a linear time invariant system from input $r_\theta(t)$ to output $\theta(t)$.

Exercise: Let $r_\theta(t) = 0$ and the initial conditions be $\theta(0) = 0.1$ rad and $\dot{\theta}(0) = 0$ rad/sec. Show that with the choice of $\alpha = \beta = 2$ the pendulum is balanced. Using MATLAB, obtain the output $\theta(t)$ for the parameters given above. Find another choice for the pair (α, β), such that $\theta(t)$ decays to zero faster without any oscillations.

2.4 Modeling Uncertainty

During the process of deriving a mathematical model for the plant, usually a series of assumptions and simplifications are made. At the end of this procedure, a *nominal plant model*, denoted by $\mathbf{P_o}$, is derived. By keeping track of the effects of simplifications and assumptions made during modeling, it is possible to derive an uncertainty description, denoted by $\mathbf{\Delta_P}$, associated with $\mathbf{P_o}$. It is then hoped (or assumed) that the true physical system lies in the set of all plants captured by the pair $(\mathbf{P_o}, \mathbf{\Delta_P})$.

2.4.1 Dynamic Uncertainty Description

Consider a nominal plant model $\mathbf{P_o}$ represented in the frequency domain by its transfer function $P_o(s)$ and suppose that the "true plant" is LTI, with unknown transfer function $P(s)$. Then the modeling uncertainty is

$$\Delta_P(s) = P(s) - P_o(s).$$

A useful *uncertainty description* in this case would be the following:

 (i) the number of poles of $P_o(s) + \Delta(s)$ in the right half plane is assumed to be the same as the number of right half plane poles of $P_o(s)$ (importance of this assumption will be clear when we discuss Nyquist stability condition and robust stability)

 (ii) also known is a function $W(s)$ whose magnitude bounds the magnitude of $\Delta_P(s)$ on the imaginary axis:

$$|\Delta_P(j\omega)| < |W(j\omega)| \quad \text{for all } \omega.$$

This type of uncertainty is called *dynamic uncertainty*. In the MIMO case, dynamic uncertainty can be structured or unstructured, in the

sense that the entries of the uncertainty matrix may or may not be independent of each other and some of the entries may be zero. The MIMO case is beyond the scope the present book, see [54] for these advanced topics and further references. For SISO plants, the pair $\{P_o(s), W(s)\}$ represents the plant model that will be used in robust controller design and analysis; see Chapter 10.

Sometimes an infinite dimensional plant model $P(s)$ is approximated by a finite dimensional model $P_o(s)$ and the difference is estimated to determine $W(s)$.

Example 2.3 Flexible Beam Model. Transfer function (2.7) of the flexible beam considered above is infinite dimensional. By taking the first few terms of the infinite product, it is possible to obtain an approximate finite dimensional model:

$$P_o(s) = \frac{2}{s^2} \prod_{n=1}^{N} \left(\frac{1 + \epsilon s - \frac{s^2}{4\alpha_n^4}}{1 + \epsilon s + \frac{s^2}{\phi_n^4}} \right).$$

Then, the difference is

$$|P(j\omega) - P_o(j\omega)| = |P_o(j\omega)| \left| 1 - \prod_{n=N+1}^{\infty} \left(\frac{1 + j\epsilon\omega + \frac{\omega^2}{4\alpha_n^4}}{1 + j\epsilon\omega - \frac{\omega^2}{\phi_n^4}} \right) \right|.$$

If N is sufficiently, large the right hand side can be bounded analytically, as demonstrated in [34].

Example 2.4 Finite Dimensional Model of a Thin Airfoil. Recall that the transfer function of a thin airfoil is in the form

$$P(s) = \frac{C_0(sI - A)^{-1}B_0}{1 - C_0(sI - A)^{-1}B_1\, \mathsf{T}(s)}$$

where $\mathsf{T}(s)$ is Theodorsen's function. By taking a finite dimensional approximation of this infinite dimensional term we obtain a finite di-

mensional plant model that is amenable for controller design:

$$P_o(s) = \frac{C_0(sI - A)^{-1}B_0}{1 - C_0(sI - A)^{-1}B_1 \, \mathsf{T}_o(s)}$$

where $\mathsf{T}_o(s)$ is a rational approximation of $\mathsf{T}(s)$. Several different approximation schemes have been studied in the literature, see for example [39], where $\mathsf{T}_o(s)$ is taken to be

$$\mathsf{T}_o(s) = \frac{(18.6 \, s_r + 1)(2.06 \, s_r + 1)}{(21.9 \, s_r + 1)(3.45 \, s_r + 1)}, \quad \text{where} \quad s_r = \frac{s \, b}{V} \, .$$

The modeling uncertainty can be bounded as follows:

$$|P(j\omega) - P_o(j\omega)| \le |P_o(j\omega)| \left| \frac{R_1(j\omega)(\mathsf{T}(j\omega) - \mathsf{T}_o(j\omega))}{1 - R_1(j\omega)\mathsf{T}(j\omega)} \right|$$

where $R_1(s) = C_0(sI - A)^{-1}B_1$. Using the bounds on approximation error, $|\mathsf{T}(j\omega) - \mathsf{T}_o(j\omega)|$, an upper bound of the right hand side can be derived; this gives $W(s)$. A numerical example can be found in [40].

2.4.2 Parametric Uncertainty Transformed to Dynamic Uncertainty

Typically, physical system parameters determine the coefficients of $P_o(s)$. Uncertain parameters lead to a special type of plant models where the structure of $P(s)$ is fixed (all possible plants $P(s)$ have the same structure as $P_o(s)$, e.g., the degrees of denominator and numerator polynomials are fixed) with uncertain coefficients. For example, consider the series RLC circuit shown in Figure 2.8, where $u(t)$ is the input voltage and $y(t)$ is the output voltage.

Transfer function of the RLC circuit is

$$P(s) = \frac{1}{LCs^2 + RCs + 1}.$$

Figure 2.8: Series RLC circuit.

So, the nominal plant model is

$$P_o(s) = \frac{1}{L_o C_o s^2 + R_o C_o s + 1},$$

where R_o, L_o, C_o are nominal values of R, L, C, respectively. Uncertainties in these parameters appear as uncertainties in the coefficients of the transfer function.

By introducing some conservatism, it is possible to transform parametric uncertainty to a dynamic uncertainty. Examples are given below.

Example 2.5 Uncertainty in damping. Consider the RLC circuit example given above. The transfer function can be rewritten as

$$P(s) = \frac{\omega_o^2}{s^2 + 2\zeta\omega_o s + \omega_o^2}$$

where $\omega_o = \frac{1}{\sqrt{LC}}$ and $\zeta = R\sqrt{C/4L}$. For the sake of argument, suppose L and C are known precisely and R is uncertain. That means ω_o is fixed and ζ varies. Consider the numerical values:

$$P(s) = \frac{1}{s^2 + 2\zeta s + 1} \qquad \zeta \in [0.1 , 0.2]$$

$$P_o(s) = \frac{1}{s^2 + 2\zeta_o s + 1} \qquad \zeta_o = 0.2.$$

Then, an uncertainty upper bound function $W(s)$ can be determined by plotting $|P(j\omega) - P_o(j\omega)|$ for a sufficiently large number of $\zeta \in [0.1 , 0.2]$.

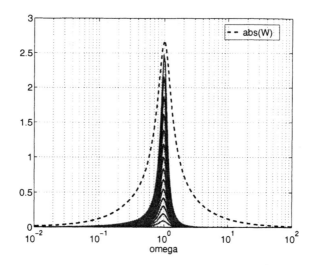

Figure 2.9: Uncertainty weight for a second order system.

The weight $W(s)$ should be such that $|W(j\omega)| > |P(j\omega) - P_o(j\omega)|$ for all P. Figure 2.9 shows that

$$W(s) = \frac{0.015\ (80\ s + 1)}{(s^2 + 0.45\ s + 1)}$$

is a feasible uncertainty weight.

Example 2.6 Uncertain time delay. A first-order stable system with time delay has transfer function in the form

$$P(s) = \frac{e^{-hs}}{(s+1)} \quad \text{where} \quad h \in [0\ , \ 0.2].$$

Suppose that time delay is ignored in the nominal model, i.e.,

$$P_o(s) = \frac{1}{(s+1)}.$$

As before, an envelop $|W(j\omega)|$ is determined by plotting the difference $|P(j\omega) - P_o(j\omega)|$ for a sufficiently large number of h between $(0\ , \ 0.2]$.

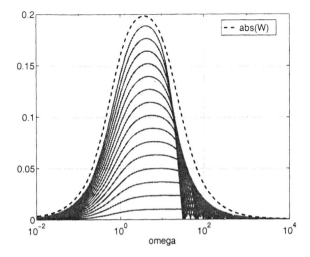

Figure 2.10: Uncertainty weight for unknown time delay.

Figure 2.10 shows that the uncertainty weight can be chosen as

$$W(s) = \frac{0.0025\,(100\ s + 1)}{(1.2\ s + 1)(0.06\ s + 1)}.$$

Note that there is conservatism here: the set

$$\mathcal{P}_h := \left\{ P(s) = \frac{e^{-hs}}{(s+1)} \ : \ h \in [0\ ,\ 0.2] \right\}$$

is a subset of

$$\mathcal{P}_\Delta := \{ P = P_o + \Delta \ : \ \Delta(s) \text{ is stable, } |\Delta(j\omega)| < |W(j\omega)| \ \forall\ \omega \},$$

which means that if a controller achieves design objectives for all plants in the set \mathcal{P}_Δ, then it is guaranteed to work for all plants in \mathcal{P}_h. Since the set \mathcal{P}_Δ is larger than the actual set of interest \mathcal{P}_h, design objectives might be more difficult to satisfy in this new setting.

2.4.3 Uncertainty from System Identification

The ultimate purpose of system identification is to derive a nominal plant model and a bound for the uncertainty. Sometimes, physical laws of nature suggest a mathematical structure for plant model (e.g., a fixed-order ordinary differential equation with unknown coefficients, such as the RLC circuit and the inverted pendulum examples given above). If the parameters of this model are unknown, they can be identified by using parameter estimation algorithms. These algorithms give estimated values of the unknown parameters, as well as bounds on the estimation errors, that can be used to determine a nominal plant model and an uncertainty bound (see [16, 32]).

In some cases, the plant is treated as a black box; assuming it is linear, an impulse (or a step) input is applied to obtain the impulse (or step) response. The data may be noisy, so the "best fit" may be an infinite dimensional model. The common practice is to find a low-order model that explains the data "reasonably well." The difference between this low-order model response and the actual output data can be seen as the response of the uncertain part of the plant. Alternatively, this difference can be treated as measurement noise whose statistical properties are to be determined.

In the black box approach, frequency domain identification techniques can also be used in finding a nominal plant/uncertainty model. For example, consider the response of a stable[1] LTI system, $P(s)$, to a sinusoidal input $u(t) = \sin(\omega_k t)$. The steady state output is

$$y_{\mathrm{ss}}(t) = |P(j\omega_k)| \sin(\omega_k t + \angle P(j\omega_k)).$$

By performing experiments for a set of frequencies $\{\omega_1, \ldots, \omega_N\}$, it is possible to obtain the frequency response data $\{P(j\omega_1), \ldots, P(j\omega_N)\}$.

[1] A precise definition of stability is given in Chapter 4, but, loosely speaking, it means that bounded inputs give rise to bounded outputs.

Note that these are complex numbers determined from steady state responses; due to measurement errors and/or unmodeled nonlinearities, there may be some uncertainty associated with each data point $P(j\omega_k)$. There are mathematical techniques to determine a nominal plant model $P_o(s)$ and an uncertainty bound $W(s)$, such that there exists a plant $P(s)$ in the set captured by the pair (P_o, W) that fits the measurements. These mathematical techniques are beyond the scope of this book; the reader is referred to [1, 25, 36] for details and further references.

2.5 Why Feedback Control?

The main reason to use feedback is to reduce the effect of "uncertainty." The uncertainty can be in the form of a modeling error in the plant description (i.e., an *unknown system*), or in the form a disturbance/noise (i.e., an *unknown signal*).

Open-loop control and feedback control schemes are compared in this section. Both the open-loop control and feedback control schemes are shown in Figure 2.11, where $r(t)$ is the reference input (i.e. desired output), $v(t)$ is the disturbance and $y(t)$ is the output. When $H = 1$, the feedback is in effect: $r(t)$ is compared with $y(t)$ and the error is fed back to the controller. Note that in a feedback system when the sensor fails (i.e., sensor output is stuck at zero) the system becomes open loop with $H = 0$.

The feedback connection $e(t) = r(t) - y(t)$ may pose a mathematical problem if the system bandwidth is infinite (i.e., both the plant and the controller are proper but not strictly proper). To see this problem, consider the trivial case where $v(t) = 0$, $P(s) = K_p$ and $C(s) = -K_p^{-1}$:

$$y(t) = -e(t) = y(t) - r(t)$$

which is meaningless for $r(t) \neq 0$. Another example is the following

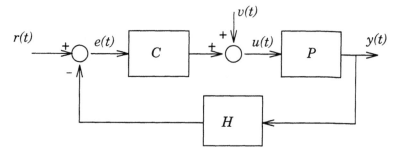

H=0 : Open Loop

H=1 : Closed Loop (feedback is in effect)

Figure 2.11: Open-loop and closed-loop systems.

situation: let $P(s) = \frac{2s}{s+2}$ and $C(s) = -0.5$, then the transfer function from $r(t)$ to $e(t)$ is

$$(1 + P(s)C(s))^{-1} = \frac{s+2}{2}$$

which is improper, i.e., non-causal, so it cannot be built physically. Generalizing the above observations, the feedback system is said to be *well-posed* if $P(\infty)C(\infty) \neq -1$. In practice, most of the physical dynamical systems do not have infinite bandwidth, i.e., $P(s)$ and hence $P(s)C(s)$ are strictly proper. So the feedback system is well posed in that case. Throughout the book, the feedback systems considered are assumed to be well-posed unless otherwise stated.

Before discussing the benefits of feedback, we should mention its obvious danger: $P(s)$ might be stable to start with, but if $C(s)$ is chosen poorly the feedback system may become unstable (i.e., a bounded reference input $r(t)$, or disturbance input $v(t)$, might lead to an unbounded signal, $u(t)$ and/or $y(t)$, within the feedback loop). In the remaining parts of this chapter, and in the next chapter, the feedback systems are assumed to be stable.

2.5.1 Disturbance Attenuation

In Figure 2.11, let $v(t) \neq 0$, and $r(t) \equiv 0$. In this situation, the magnitude of the output, $y(t)$, should be as small as possible so that it is as close to desired response, $r(t)$, as possible.

For the open-loop control scheme, $H = 0$, the output is

$$Y(s) = P(s)(V(s) + C(s)R(s)).$$

Since $R(s) = 0$ the controller does not play a role in the disturbance response, $Y(s) = P(s)V(s)$. When $H = 1$, the feedback is in effect; in this case

$$Y(s) = \frac{P(s)}{1 + P(s)C(s)}(V(s) + C(s)R(s)).$$

So the response due to $v(t)$ is $Y(s) = P(s)(1 + P(s)C(s))^{-1}V(s)$. Note that the closed-loop response is equal to the open-loop response multiplied by the factor $(1 + P(s)C(s))^{-1}$. For good disturbance attenuation we need to make this factor small by an appropriate choice of $C(s)$.

Let $|V(j\omega)|$ be the magnitude of the disturbance in frequency domain, and, for the sake of argument, suppose that $|V(j\omega)| \neq 0$ for $\omega \in \Omega$, and $|V(j\omega)| \approx 0$ for ω outside the frequency region defined by Ω. If the controller is designed in such a way that

$$|(1 + P(j\omega)C(j\omega))^{-1}| \ll 1 \quad \forall \ \omega \in \Omega \qquad (2.13)$$

then high attenuation is achieved by feedback. The disturbance attenuation factor is the left hand side of (2.13).

2.5.2 Tracking

Now consider the dual problem where $r(t) \neq 0$, and $v(t) \equiv 0$. In this case, *tracking error*, $e(t) := r(t) - y(t)$ should be as small as possible.

In the open-loop case, the goal is achieved if $C(s) = 1/P(s)$. But note that if $P(s)$ is strictly proper, then $C(s)$ is improper, i.e., non-causal. To avoid this problem, one might approximate $1/P(s)$ in the region of the complex plane where $|R(s)|$ is large. But if $P(s)$ is unstable and if there is uncertainty in the right half plane pole location, then $1/P(s)$ cannot be implemented precisely, and the tracking error is unbounded.

In the feedback scheme, the tracking error is

$$E(s) = (1 + P(s)C(s))^{-1} R(s).$$

Therefore, similar to disturbance attenuation, one should select $C(s)$ in such a way that $|(1 + P(j\omega)C(j\omega))^{-1}| \ll 1$ in the frequency region where $|R(j\omega)|$ is large.

2.5.3 Sensitivity to Plant Uncertainty

For a function F, which depends on a parameter α, sensitivity of F to variations in α is denoted by S_α^F, and it is defined as follows

$$S_\alpha^F := \lim_{\Delta_\alpha \to 0} \left. \frac{\Delta_F/F}{\Delta_\alpha/\alpha} \right|_{\alpha=\alpha_o} = \left. \frac{\alpha}{F} \frac{\partial F}{\partial \alpha} \right|_{\alpha=\alpha_o}$$

where α_o is the nominal value of α, Δ_α and Δ_F represent the deviations of α and F from their nominal values α_o and F evaluated at α_o, respectively.

Transfer function from reference input $r(t)$ to output $y(t)$ is

$$T_{ol}(s) = P(s)C(s) \quad \text{(for an open-loop system)},$$

$$T_{cl}(s) = \frac{P(s)C(s)}{1 + P(s)C(s)} \quad \text{(for a closed-loop system)}.$$

Typically the plant is uncertain, so it is in the form $P = P_o + \Delta_P$. Then the above transfer functions can be written as $T_{ol} = T_{ol,o} + \Delta_{T_{ol}}$ and

$T_{cl} = T_{cl,o} + \Delta_{T_{cl}}$, where $T_{ol,o}$ and $T_{cl,o}$ are the nominal values when P is replaced by P_o. Applying the definition, sensitivities of T_{ol} and T_{cl} to variations in P are

$$S_P^{T_{ol}} = \lim_{\Delta P \to 0} \frac{\Delta_{T_{ol}}/T_{ol,o}}{\Delta P/P_o} = 1 \qquad (2.14)$$

$$S_P^{T_{cl}} = \lim_{\Delta P \to 0} \frac{\Delta_{T_{cl}}/T_{cl,o}}{\Delta P/P_o} = \frac{1}{1 + P_o(s)C(s)}. \qquad (2.15)$$

The first equation (2.14) means that the percentage change in T_{ol} is equal to the the percentage change in P. The second equation (2.15) implies that if there is a frequency region where percentage variations in T_{cl} should be made small, then the controller can be chosen in such a way that the function $(1 + P_o(s)C(s))^{-1}$ has small magnitude in that frequency range. Hence, the effect of variations in P can be made small by using feedback control; the same cannot be achieved by open-loop control.

In the light of (2.15) the function $(1 + P_o(s)C(s))^{-1}$ is called the "nominal sensitivity function" and it is denoted by $S(s)$. The sensitivity function, denoted by $S_\Delta(s)$, is the same function when P_o is replaced by $P = P_o + \Delta_P$. In all the examples seen above, sensitivity function plays an important role. One of the most important design goals in feedback control is sensitivity minimization. This is discussed further in Chapter 10.

2.6 Exercise Problems

1. Consider the flow control problem illustrated in Figure 2.2, and assume that the outgoing flow rate $v(t)$ is proportional to the square root of the liquid level $h(t)$ in the reservoir (e.g., this is the case if the liquid flows out through a valve with constant opening):

$$v(t) = v_0 \sqrt{h(t)}.$$

Furthermore, suppose that the area of the reservoir $A(h(t))$ is constant, say A_0. Since total accumulation is $y(t) = A_0 h(t)$, dynamical equation for this system is

$$\dot{h}(t) = \frac{1}{A_0} \left(u(t - \tau) - v_0 \sqrt{h(t)} \right).$$

Let $h(t) = h_0 + \delta_h(t)$ and $u(t) = u_0 + \delta_u(t)$, with $u_0 = v_0 \sqrt{h_0}$. Linearize the system around the operating point h_0 and find the transfer function from $\delta_u(t)$ to $\delta_h(t)$.

2. For the above mentioned flow control problem, suppose that the geometry of the reservoir is known as

$$A(h(t)) = A_0 + A_1 h(t) + A_2 \sqrt{h(t)}$$

with some constants A_0, A_1, A_2. Let the outgoing flow rate be $v(t) = v_0 + w(t)$, with $v_0 > |w(t)| \geq 0$. The term $w(t)$ can be seen as a disturbance representing the "load variations" around the nominal constant load v_0; typically $w(t)$ is a superposition of a finite number of sine and cosine functions.

(i) Assume that $\tau = 0$ and $v(t)$ is available to the controller. Given a desired liquid level $h_d(t)$, there exists a feedback control input $u(t)$ (a nonlinear function of $h(t)$, $v(t)$ and $h_d(t)$) linearizing the system whose input is $h_d(t)$ and output is $h(t)$. For $h_d(t) = h_d$, find such $u(t)$ that leads to $h(t) \to h_d$ as $t \to \infty$ for any initial condition $h(0) = h_0$.

(ii) Now consider a linear controller in the form

$$u(t) = K \left(h_d(t) - h(t) \right) + v_0$$

(in this case, the controller does not have access to $w(t)$), where the gain $K > 0$ is to be determined. Let

$$A_0 = 300, \quad A_1 = 12, \quad A_2 = 120, \quad \tau = 12$$
$$w(t) = 20 \sin(0.5t), \quad v_0 = 24, \quad h_d(t) \equiv 10, \quad h(0) = 8,$$

(note that time delay is non-zero in this case). By using Euler's method, simulate the feedback system response for several different values of $K \in [10 , 110]$. Find a value of K for which

$$|h_d - h(t)| \leq 0.05 \quad \forall \quad t \geq 50 .$$

Note: to simulate a nonlinear system in the form

$$\dot{x}(t) = f(x(t), t)$$

first select equally spaced time instants $t_k = kT_s$, for some $T_s = (t_{k+1} - t_k) > 0$, $k \geq 0$. For example, in the above problem, T_s can be chosen as $T_s \approx 0.04$. Then, given $x(t_0)$, we can determine $x(t_k)$ from $x(t_{k-1})$ as follows:

$$x(t_k) \cong x(t_{k-1}) + T_s \ f(x(t_{k-1}), t_{k-1}) \quad \text{for } k \geq 1.$$

This approximation scheme is called Euler's method. More accurate and sophisticated simulation techniques are available; these, as well as the Euler's method, are implemented in the SIMULINK package of MATLAB.

3. An RLC circuit has transfer function in the form

$$P(s) = \frac{\omega_o^2}{s^2 + 2\zeta\omega_o s + \omega_o^2} .$$

Let $\zeta_n = 0.2$ and $\omega_{o,n} = 10$ be the nominal values of the parameters of $P(s)$, and determine the poles of $P_o(s)$.

(i) Find an uncertainty bound $W(s)$ for $\pm 20\%$ uncertainty in the values of ζ and ω_o.

(ii) Determine the sensitivity of $P(s)$ to variations in ζ.

4. For the flexible beam model, to obtain a nominal finite dimensional transfer function take $N = 4$ and determine $P_o(s)$ by computing α_n and ϕ_n, for $n = 1, \ldots, 4$. What are the poles and zeros of $P_o(s)$? Plot the difference $|P(j\omega) - P_o(j\omega)|$ by writing a

MATLAB script, and find a low-order (at most 2nd-order) rational function $W(s)$ that is a feasible uncertainty bound.

5. Consider the disturbance attenuation problem for a first-order plant $P(s) = 20/(s + 4)$ with $v(t) = \sin(3t)$. For the open-loop system, magnitude of the steady state output is $|P(j3)| = 4$ (i.e., the disturbance is amplified). Show that in the feedback scheme a necessary condition for the steady state output to be zero is $|C(\pm j3)| = \infty$ (i.e. the controller must have a pair of poles at $s = \pm j3$).

Chapter 3

Performance Objectives

Basic principles of feedback control are discussed in the previous chapter. We have seen that the most important role of the feedback is to reduce the effects of uncertainty. In this chapter, time domain performance objectives are defined for certain special tracking problems. Plant uncertainty and disturbances are neglected in this discussion.

3.1 Step Response: Transient Analysis

In the standard feedback control system shown in Figure 2.11, assume that the transfer function from $r(t)$ to $y(t)$ is in the form

$$T(s) = \frac{\omega_o^2}{s^2 + 2\zeta\omega_o s + \omega_o^2} \qquad 0 < \zeta < 1 \qquad \omega_o \in \mathbb{R}$$

and $r(t)$ is the unit step function, denoted by $\mathbb{U}(t)$. Then, the output $y(t)$ is the inverse Laplace transform of

$$Y(s) = \frac{\omega_o^2}{(s^2 + 2\zeta\omega_o s + \omega_o^2)} \frac{1}{s}$$

that is

$$y(t) = 1 - \frac{e^{-\zeta \omega_o t}}{\sqrt{1 - \zeta^2}} \sin(\omega_d t + \theta) \quad t \geq 0,$$

where $\omega_d := \omega_o \sqrt{1 - \zeta^2}$ and $\theta := \cos^{-1}(\zeta)$. For some typical values of ζ, the step response $y(t)$ is as shown in Figure 3.1. Note that the steady state value of $y(t)$ is $y_{ss} = 1$ because $T(0) = 1$. Steady state response is discussed in more detail in the next section.

The maximum *percent overshoot* is defined to be the quantity

$$PO := \frac{y_p - y_{ss}}{y_{ss}} \times 100\%$$

where y_p is the peak value. By simple calculations it can be seen that the peak value of $y(t)$ occurs at the time instant $t_p = \pi/\omega_d$, and

$$PO = e^{-\pi \zeta / \sqrt{1 - \zeta^2}} \times 100\%.$$

Figure 3.2 shows PO versus ζ. Note that the output is desired to reach its steady state value as fast as possible with a reasonably small PO. In order to have a small PO, ζ should be large. For example, if $PO \leq 10\%$ is desired, then ζ must be greater or equal to 0.6.

The *settling time* is defined to be the smallest time instant t_s, after which the response $y(t)$ remains within 2% of its final value, i.e.,

$$t_s := \min\{ t' : |y(t) - y_{ss}| \leq 0.02 \, y_{ss} \quad \forall \, t \geq t'\}.$$

Sometimes 1% or 5% is used in the definition of settling time instead of 2%. Conceptually, they are not significantly different. For the second-order system response, with the 2% definition of the settling time,

$$t_s \approx \frac{4}{\zeta \omega_o}.$$

So, in order to have a fast settling response, the product $\zeta \omega_o$ should be large.

Figure 3.1: Step response of a second-order system.

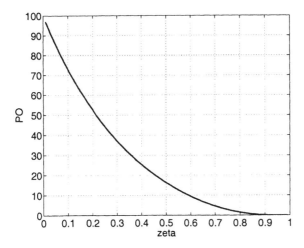

Figure 3.2: PO versus ζ.

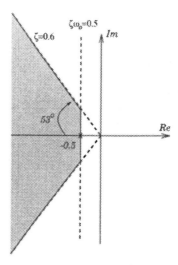

Figure 3.3: Region of the desired closed-loop poles.

The poles of $T(s)$ are

$$r_{1,2} = -\zeta\omega_o \pm j\omega_o\sqrt{1-\zeta^2}.$$

Therefore, once the maximum allowable settling time and PO are specified, the poles of $T(s)$ should lie in a region of the complex plane defined by minimum allowable ζ and $\zeta\omega_o$.

For example, let the desired PO and t_s be bounded by

$$\text{PO} \leq 10\% \qquad \text{and} \qquad t_s \leq 8 \text{ sec}.$$

For these design specifications, the region of the complex plane in which closed-loop system poles should lie is determined as follows. The PO requirement implies that $\zeta \geq 0.6$, equivalently $\theta \leq 53°$, (recall that $\cos(\theta) = \zeta$). The settling time requirement is satisfied if and only if $\text{Re}(r_{1,2}) \leq -0.5$. Then, the region of desired closed-loop poles is the shaded area shown in Figure 3.3.

If the order of the closed-loop transfer function $T(s)$ is higher than two, then, depending on the location of its poles and zeros, it may be possible to approximate the closed-loop step response by the response of a second-order system. For example, consider the third-order system

$$T(s) = \frac{\omega_o^2}{(s^2 + 2\zeta\omega_o s + \omega_o^2)(1 + s/r)} \qquad \text{where} \quad r \gg \zeta\omega_o.$$

The transient response contains a term e^{-rt}. Compared with the envelope $e^{-\zeta\omega_o t}$ of the sinusoidal term, e^{-rt} decays very fast, and the overall response is similar to the response of a second-order system. Hence, the effect of the third pole $r_3 = -r$ is negligible. Consider another example,

$$T(s) = \frac{\omega_o^2(1 + s/(r + \epsilon))}{(s^2 + 2\zeta\omega_o s + \omega_o^2)(1 + s/r)} \qquad \text{where} \quad 0 < \epsilon \ll r.$$

In this case, although r does not need to be much larger than $\zeta\omega_o$, the zero at $-(r+\epsilon)$ cancels the effect of the pole at $-r$. To see this, consider the partial fraction expansion of $Y(s) = T(s)R(s)$ with $R(s) = 1/s$

$$Y(s) = \frac{A_0}{s} + \frac{A_1}{s - r_1} + \frac{A_2}{s - r_2} + \frac{A_3}{s + r} \qquad \text{where} \quad A_0 = 1 \text{ and}$$

$$A_3 = \lim_{s \to -r}(s + r)Y(s) = \frac{\omega_o^2}{2\zeta\omega_o r - (\omega_o^2 + r^2)}\left(\frac{\epsilon}{r + \epsilon}\right).$$

Since $|A_3| \to 0$ as $\epsilon \mapsto 0$, the term $A_3 e^{-rt}$ is negligible in $y(t)$.

In summary, if there is an approximate pole zero cancelation in the left half plane, then this pole-zero pair can be taken out of the transfer function $T(s)$ to determine PO and t_s. Also, the poles closest to the imaginary axis dominate the transient response of $y(t)$. To generalize this observation, let r_1, \ldots, r_n be the poles of $T(s)$, such that $\text{Re}(r_k) \ll \text{Re}(r_2) = \text{Re}(r_1) < 0$, for all $k \geq 3$. Then, the pair of complex conjugate poles $r_{1,2}$ are called the *dominant poles*. We have seen that the desired transient response properties, e.g., PO and t_s, can be translated into requirements on the location of the dominant poles.

3.2 Steady State Analysis

For the standard feedback system of Figure 2.11, the tracking error is defined to be the signal $e(t) := r(t) - y(t)$. One of the typical performance objectives is to keep the magnitude of the steady state error,

$$e_{ss} := \lim_{t \to \infty} e(t), \tag{3.1}$$

within a specified bound. Whenever the above limit exists (i.e., $e(t)$ converges) the final value theorem can be applied:

$$e_{ss} = \lim_{s \to 0} sE(s) = \lim_{s \to 0} s \frac{1}{1 + G(s)} R(s),$$

where $G(s) = P(s)C(s)$ is the open-loop transfer function from r to y. Suppose that $G(s)$ is in the form

$$G(s) = \frac{N_G(s)}{s^\ell \tilde{D}_G(s)} \tag{3.2}$$

with $\ell \geq 0$ and $\tilde{D}_G(0) \neq 0 \neq N_G(0)$. Let the reference input be

$$r(t) = t^{k-1} \mathbf{U}(t) \longleftrightarrow R(s) = \frac{1}{s^k} \quad k \geq 1.$$

When $k = 1$ (i.e. $r(t)$ is unit step) the steady state error is

$$e_{ss} = \begin{cases} (1 + G(0))^{-1} & \text{if } \ell = 0 \\ 0 & \text{if } \ell \geq 1. \end{cases}$$

If zero steady state error is desired for unit step reference input, then $G(s) = P(s)C(s)$ must have at least one pole at $s = 0$. For $k \geq 2$

$$e_{ss} = \begin{cases} \infty & \text{if } \ell \leq k - 2 \\ \frac{\tilde{D}_G(0)}{N_G(0)} & \text{if } \ell = k - 1 \\ 0 & \text{if } \ell \geq k. \end{cases}$$

The system is said to be *type k* if $e_{ss} = 0$ for $R(s) = 1/s^k$. For the standard feedback system where $G(s)$ is in the form (3.2) the system is type k only if $\ell \geq k$.

Now consider a sinusoidal reference input $R(s) = \frac{\omega_o^2}{s^2 + \omega_o^2}$. The tracking error $e(t)$ is the inverse Laplace transform of

$$E(s) = \frac{1}{1 + G(s)} \left(\frac{\omega_o^2}{s^2 + \omega_o^2} \right).$$

Partial fraction expansion and inverse Laplace transformation yield a sinusoidal term, $A_o \sin(\omega_o t + \phi)$, in $e(t)$. Unless $A_o = 0$, the limit (3.1) does not exist, hence the final value theorem cannot be applied. In order to have $A_o = 0$, the transfer function $S(s) = (1 + G(s))^{-1}$ must have zeros at $s = \pm j\omega_o$, i.e. $G(s)$ must be in the form

$$G(s) = \frac{N_G(s)}{(s^2 + \omega_o^2)\widehat{D}_G(s)} \qquad \text{where } N_G(\pm j\omega_o) \neq 0.$$

In conclusion, $e_{ss} = 0$ only if the zeros of $S(s)$ (equivalently, the poles of $P(s)C(s) = G(s)$) include all Im-axis poles of $R(s)$. Let η_1, \ldots, η_k be the Im-axis poles of $R(s)$, including multiplicities. Assuming that none of the η_i's are poles of $P(s)$, to achieve $e_{ss} = 0$ the controller must be in form

$$C(s) = \widetilde{C}(s) \frac{1}{D_R(s)} \tag{3.3}$$

where $D_R(s) := (s - \eta_1) \ldots (s - \eta_k)$, and $\widetilde{C}(\eta_i) \neq 0$ for $i = 1, \ldots, k$.

For example, when $R(s) = 1/s$ and $|P(0)| < \infty$, we need a pole at $s = 0$ in the controller to have $e_{ss} = 0$. A simple example is PID (proportional, integral, derivative) controller, which is in the form

$$C(s) = \left(K_p + \frac{K_i}{s} + K_d s \right) \left(\frac{1}{1 + \epsilon s} \right) \tag{3.4}$$

where K_p, K_i and K_d are the proportional, integral, and derivative action coefficients, respectively; the term $\epsilon \searrow 0$ is needed to make the

controller proper. When $K_d = 0$ we can set $\epsilon = 0$; in this case $C(s)$, (3.4), becomes a PI controller. See Section 8.4 for further discussions on PID controllers.

Note that a copy of the reference signal generator, $\frac{1}{D_R(s)}$, is included in the controller (3.3). We can think of $\tilde{C}(s)$ as the "controller" to be designed for the "plant"

$$\tilde{P}(s) := P(s) \, \frac{1}{D_R(s)}.$$

This idea is the basis of *servocompensator* design, [14, 21] and *repetitive controller* design, [24, 43].

3.3 Exercise Problems

1. Consider the second-order system with a zero

$$T(s) = \frac{\omega_o^2 \, (1 - s/z)}{s^2 + 2\zeta\omega_o s + \omega_o^2}$$

where $\zeta \in (0 , 1)$ and $z \in \mathbb{R}$. What is the steady state error for unit step reference input? Let $\omega_o = 1$ and plot the step response for $\zeta = 0.1 , 0.5, 0.9$ and $z = -5, -1, 0.2, 3$.

2. For $P(s) = 1/s$ design a controller in the form

$$C(s) = \frac{K_c(s + z_c)^2}{s^2 + \omega_o^2}$$

so that the sensitivity function $S(s) = (1 + P(s)C(s))^{-1}$ has three zeros at $0, \pm j\omega_o$ and three poles with real parts less than -0.5. This guarantees that $e_{ss} = 0$ for reference inputs $r(t) = \mathbb{U}(t)$ and $r(t) = \sin(2t)\mathbb{U}(t)$. Plot the closed-loop system response corresponding to these reference inputs for your design.

Chapter 4

BIBO Stability

In this chapter, bounded input–bounded output (BIBO) stability of a linear time invariant (LTI) system will be defined first. Then, we will see that BIBO stability of the *feedback system* formed by a controller $C(s)$ and a plant $P(s)$ can determined by applying the Routh-Hurwitz test on the *characteristic polynomial*, which is defined from the numerator and denominator polynomials of $C(s)$ and $P(s)$. For systems with parametric uncertainty in the coefficients of the transfer function, we will see Kharitanov's robust stability test and its extensions.

4.1 Norms for Signals and Systems

In system analysis, input and output signals are considered as functions of time. For example $u(t)$, and $y(t)$, for $t \geq 0$, are input and output functions of the LTI system \mathbf{F}, shown in Figure 4.1.

Each of these time functions is assumed to be an element of a function space, $u \in \mathcal{U}$, and $y \in \mathcal{Y}$, where \mathcal{U} represents the set of possible input functions, and similarly \mathcal{Y} represents the set of output functions.

Figure 4.1: Linear time invariant system.

For mathematical convenience, \mathcal{U} and \mathcal{Y} are usually taken to be vector spaces on which signal norms are defined. The norm $\|u\|_{\mathcal{U}}$ is a measure of how large the input signal is, similarly for the output signal norm $\|y\|_{\mathcal{Y}}$. Using this abstract notation we can define the system norm, denoted by $\|\mathbf{F}\|$, as the quantity

$$\|\mathbf{F}\| = \sup_{u \neq 0} \frac{\|y\|_{\mathcal{Y}}}{\|u\|_{\mathcal{U}}}. \tag{4.1}$$

A physical interpretation of (4.1) is the following: the ratio $\frac{\|y\|_{\mathcal{Y}}}{\|u\|_{\mathcal{U}}}$ represents the amplification factor of the system for a fixed input $u \neq 0$. Since the largest possible ratio is taken (sup means the least upper bound) as the system norm, $\|\mathbf{F}\|$ can be interpreted as the largest signal amplification through the system \mathbf{F}.

In signals and systems theory, most widely used function spaces are $\mathcal{L}_1[0, \infty)$, $\mathcal{L}_2[0, \infty)$, and $\mathcal{L}_\infty[0, \infty)$. Precise definitions of these Lebesgue spaces are beyond the scope of this book. They can be loosely defined as follows: for $1 \leq p < \infty$

$$\mathcal{L}_p[0, \infty) = \left\{ f : [0, \infty) \to \mathbb{R}, \ \|f\|_{\mathcal{L}_p}^p := \int_0^\infty |f(t)|^p dt < \infty \right\}$$

$$\mathcal{L}_\infty[0, \infty) = \left\{ f : [0, \infty) \to \mathbb{R}, \ \|f\|_{\mathcal{L}_\infty} := \sup_{t \in [0, \infty)} |f(t)| < \infty \right\}.$$

Note that $\mathcal{L}_2[0, \infty)$ is the space of all finite energy signals, and $\mathcal{L}_\infty[0, \infty)$ is the set of all bounded signals. In the above definitions the real valued function $f(t)$ is defined on the positive time axis and it is assumed to be piecewise continuous. An impulse, e.g., $\delta(t - t_o)$ for some $t_o \geq 0$, does not belong to any of these function spaces. So, it is useful to define a

space of impulse functions,

$$\mathcal{I}_1 \;\; := \;\; \left\{ f(t) = \sum_{k=1}^{\infty} \alpha_k \delta(t - t_k), \;\; t_k \geq 0, \;\; \alpha_k \in \mathbb{R}, \;\; \sum_{k=1}^{\infty} |\alpha_k| < \infty \right\}$$

$$\text{with the norm defined as } \; \|f\|_{\mathcal{I}_1} := \sum_{k=1}^{\infty} |\alpha_k|.$$

The functions from $\mathcal{L}_1[0, \infty)$ and \mathcal{I}_1 can be combined to obtain another function space

$$\mathcal{A}_1 := \{ f(t) = g(t) + h(t) \;\; : \;\; g \in \mathcal{L}_1[0, \infty) \, , \;\; h \in \mathcal{I}_1 \} .$$

For example,

$$f(t) = 3e^{-2t} \sin(10\pi t) - 4\delta(t - \pi) + 5\delta(t - 10) \, , \quad t \geq 0,$$

belongs to \mathcal{A}_1, but it does not belong to $\mathcal{L}_2[0, \infty)$, nor $\mathcal{L}_\infty[0, \infty)$. The importance of \mathcal{A}_1 will be clear shortly.

Exercise. Determine whether time functions given below belong to any of the function spaces defined above:

$$f_1(t) \;\; = \;\; (t + 1)^{-1} \, , \quad f_2(t) = (t + 1)^{-2} \, , \quad f_3(t) \frac{3}{\sqrt{t + 1}}$$

$$f_4(t) \;\; = \;\; \frac{\sin(4\pi t)}{t} \, , \quad f_5(t) = (t - 1)^{-1/5} \, , \quad f_6(t) = (t - 1)^{-6}.$$

4.2 BIBO Stability

Formally, a system \mathbf{F} is said to be *bounded input–bounded output (BIBO) stable* if every bounded input u generates a bounded output y, and the largest signal amplification through the system is finite. In the abstract notation, that means BIBO stability is equivalent to having a finite system norm, i.e., $\|\mathbf{F}\| < \infty$. Note that definition of BIBO stability depends on the selection of input and output spaces. The most common

definition of BIBO stability deals with the special case where the input and output spaces are $\mathcal{U} = \mathcal{Y} = \mathcal{L}_\infty[0, \infty)$. Let $f(t)$ be the impulse response and $F(s)$ be the transfer function (i.e., the Laplace transform of $f(t)$) of a causal LTI system \mathbf{F}.

Theorem 4.1 *Suppose $\mathcal{U} = \mathcal{Y} = \mathcal{L}_\infty[0, \infty)$. Then the system \mathbf{F} is BIBO stable if and only if $f \in \mathcal{A}_1$. Moreover,*

$$\|\mathbf{F}\| = \|f\|_{\mathcal{A}_1}. \tag{4.2}$$

Proof. The result follows from the convolution identity

$$y(t) = \int_0^\infty f(\tau)u(t - \tau)d\tau. \tag{4.3}$$

If $u \in \mathcal{L}_\infty[0, \infty)$, then $|u(t - \tau)| \leq \|u\|_{\mathcal{L}_\infty}$ for all t and τ. It is easy to verify the following inequalities:

$$
\begin{aligned}
|y(t)| &= \left| \int_0^\infty f(\tau)u(t - \tau)d\tau \right| \\
&\leq \int_0^\infty |f(\tau)|\,|u(t - \tau)|d\tau \\
&\leq \|u\|_{\mathcal{L}_\infty} \int_0^\infty |f(\tau)|d\tau \;=\; \|u\|_{\mathcal{L}_\infty}\|f\|_{\mathcal{A}_1}.
\end{aligned}
$$

Hence, for all $u \neq 0$

$$\frac{\|y\|_{\mathcal{L}_\infty}}{\|u\|_{\mathcal{L}_\infty}} \leq \|f\|_{\mathcal{A}_1}$$

which means that $\|\mathbf{F}\| \leq \|f\|_{\mathcal{A}_1}$. In order to prove the converse, consider (4.3) in the limit as $t \to \infty$ with u defined at time instant τ as

$$
u(t - \tau) = \begin{cases} 1 & \text{if } f(\tau) > 0, \\ -1 & \text{if } f(\tau) < 0, \\ 0 & \text{if } f(\tau) = 0. \end{cases}
$$

Clearly $|y(t)| \to \|f\|_{A_1}$ in the limit as $t \to \infty$. Thus

$$\|\mathbf{F}\| = \sup_{u \neq 0} \frac{\|y\|_{\mathcal{L}_\infty}}{\|u\|_{\mathcal{L}_\infty}} \geq \|f\|_{A_1}.$$

This concludes the proof.

Exercise: Determine BIBO stability of the LTI system whose impulse response is (a) $f_0(t)$, (b) $f_1(t)$, where

$$f_0(t) := \begin{cases} 0 & \text{if } t < 0 \\ 1 & \text{if } 0 \leq t \leq t_0 \\ (t-1)^{-k} & \text{if } t > t_0 \end{cases} \qquad f_1(t) := \begin{cases} 0 & \text{if } t < 0 \\ t^{-\ell} & \text{if } 0 \leq t \leq t_1 \\ 0 & \text{if } t > t_1 \end{cases}$$

with $t_0 > 1$, $k > 1$, and $t_1 > 0$, $0 < \ell < 1$.

Theorem 4.2 *Suppose* $\mathcal{U} = \mathcal{Y} = \mathcal{L}_2[0, \infty)$. *Then the system norm of* **F** *is*

$$\|\mathbf{F}\| = \sup_{\sigma > 0,\ \omega \in \mathbb{R}} |F(\sigma + j\omega)| =: \|F\|_{\mathcal{H}^\infty} \qquad (4.4)$$

That is the system **F** *is stable if and only if its transfer function has no poles in* $\overline{\mathbb{C}}_+$. *Moreover, when the system is stable, maximum modulus principle implies that*

$$\|F\|_{\mathcal{H}^\infty} = \sup_{\omega \in \mathbb{R}} |F(j\omega)| =: \|F\|_{\mathcal{L}^\infty}. \qquad (4.5)$$

See, e.g., [18, pp. 16–17] and [20, pp. 18–20] for proofs of this theorem. Further discussions can also be found in [15]. The proof is based on Parseval's identity, which says that the energy of a signal can be computed from its Fourier transform. The equivalence (4.5) implies that when the system is stable, its norm (in the sense of largest energy amplification) can be computed from the peak value of its Bode magnitude plot. The second equality in (4.4) implicitly defines the space \mathcal{H}^∞ as

the set of all analytic functions of s (the Laplace transform variable) that are bounded in the right half plane \mathbb{C}_+.

In this book, we will mostly consider systems with real rational transfer functions of the form $F(s) = N_F(s)/D_F(s)$ where $N_F(s)$ and $D_F(s)$ are polynomials in s with real coefficients. For such systems the following holds

$$\|f\|_{\mathcal{A}_1} < \infty \iff \|F\|_{\mathcal{H}^\infty} < \infty. \tag{4.6}$$

So, stability tests resulting from (4.2) and (4.4) are equivalent for this class of system. To see the equivalence (4.6) we can rewrite $F(s)$ in the form of partial fraction expansions and use the Laplace transform identities for each term to obtain $f(t)$ in the form

$$f(t) = \sum_{k=1}^{n} \sum_{\ell=0}^{m_k-1} c_{k\ell} \, t^\ell \, e^{p_k t} \tag{4.7}$$

where p_1, \ldots, p_n are distinct poles of $F(s)$ with multiplicities m_1, \ldots, m_n, respectively, and $c_{k\ell}$ are constant coefficients. The total number of poles of $F(s)$ is $(m_1 + \cdots + m_n)$. It is now clear that $\|f\|_{\mathcal{A}_1}$ is finite if and only if $p_k \in \mathbb{C}_-$, i.e. $\mathrm{Re}(p_k) < 0$ for all $k = 1, \ldots, n$, and this condition holds if and only if $F \in \mathcal{H}^\infty$.

Rational transfer functions are widely used in control engineering practice. However, they do not capture spatially distributed parameter systems (e.g., flexible beams) and systems with time delays. Later in the book, systems with simple time delays will also be considered. The class of delay systems we will be dealing with have transfer functions in the form

$$F(s) = \frac{N_0(s) + e^{-\tau_n s} N_1(s)}{D_0(s) + e^{-\tau_d s} D_1(s)}$$

where $\tau_n \geq 0$, $\tau_d \geq 0$, and N_0, N_1, D_0, D_1 are polynomials with real coefficients, satisfying $\deg(D_0) > \max\{\deg(N_0), \deg(N_1), \deg(D_1)\}$. The

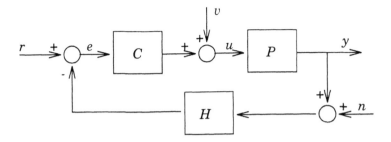

Figure 4.2: Feedback system.

equivalence (4.6) holds for this type of systems as well. Although there may be infinitely many poles in this case, the number of poles to the right of any vertical axis in the complex plane is finite (see Chapter 7).

In summary, for the class of systems we consider in this book, a system \mathbf{F} is stable if and only if its transfer function $F(s)$ is bounded and analytic in $\overline{\mathbb{C}}_+$, i.e. $F(s)$ does not have any poles in $\overline{\mathbb{C}}_+$.

4.3 Feedback System Stability

In the above section, stability of a causal single-input–single-output (SISO) LTI system is discussed. Definition of stability can be easily extended to multi-input–multi-output (MIMO) LTI systems as follows. Let $u_1(t), \ldots, u_k(t)$ be the inputs and $y_1(t), \ldots, y_\ell(t)$ be the outputs of such a system \mathbf{F}. Define $F_{ij}(s)$ to be the transfer function from u_j to y_i, $i = 1, \ldots, \ell$, and $j = 1, \ldots, k$. Then, \mathbf{F} is stable if and only if each $F_{ij}(s)$ is a stable transfer function (i.e., it has no poles in $\overline{\mathbb{C}}_+$) for all $i = 1, \ldots, \ell$, and $j = 1, \ldots, k$.

The standard feedback system shown in Figure 4.2 can be considered as a single MIMO system with inputs $r(t)$, $v(t)$, $n(t)$ and outputs $e(t)$, $u(t)$, $y(t)$. The feedback system is stable if and only if all closed-loop transfer functions are stable. Let the closed-loop transfer function from

input $r(t)$ to output $e(t)$ be denoted by $T_{re}(s)$, and similarly for the remaining eight closed-loop transfer functions. It is an easy exercise to verify that

$$
\begin{aligned}
T_{re} &= S & T_{ve} &= -HPS & T_{ne} &= -HS \\
T_{ru} &= CS & T_{vu} &= S & T_{nu} &= -HCS \\
T_{ry} &= PCS & T_{vy} &= PS & T_{ny} &= -HPCS
\end{aligned}
$$

where dependence on s is dropped for notational convenience, and

$$
S(s) := (1 + H(s)P(s)C(s))^{-1}.
$$

In this configuration, $P(s)$ and $H(s)$ are given and $C(s)$ is to be designed. The primary design goal is closed-loop system stability. In engineering applications, the sensor model $H(s)$ is usually a stable transfer function. Depending on the measurement setup, $H(s)$ may be non-minimum phase. For example, this is the case if the actual plant output is measured indirectly with a certain time delay. The plant $P(s)$ may or may not be stable. If $P(s)$ is unstable, none of its poles in $\overline{\mathbb{C}}_+$ should coincide a zero of $H(s)$. Otherwise, it is impossible to stabilize the feedback system because in this case T_{vy} and T_{ry} are unstable independent of $C(s)$, though $S(s)$ may be stable. Similarly, it is easy to show that if there is a pole zero cancelation in $\overline{\mathbb{C}}_+$ in the product $H(s)P(s)C(s)$, then one of the closed-loop transfer functions is unstable. For example, let $H(s) = 1$, $P(s) = (s-1)(s+2)^{-1}$ and $C(s) = (s-1)^{-1}$. In this case, T_{ru} is unstable.

In the light of the above discussion, suppose there is no unstable pole-zero cancelation in the product $H(s)P(s)C(s)$. Then the feedback system is stable if and only if the roots of

$$
1 + H(s)P(s)C(s) = 0 \tag{4.8}
$$

are in \mathbb{C}_-. For the purpose of investigating closed-loop stability and controller design, we can define $P_H(s) := H(s)P(s)$ as "the plant seen

by the controller." Therefore, without loss of generality, we will assume that $H(s) = 1$ and $P_H(s) = P(s)$.

Now consider the finite dimensional case where $P(s)$ and $C(s)$ are rational functions, i.e., there exist polynomials $N_P(s)$, $D_P(s)$, $N_C(s)$ and $D_C(s)$ such that $P(s) = N_P(s)/D_P(s)$, $C(s) = N_C(s)/D_C(s)$ and

(N_P, D_P) and (N_C, D_C) are coprime pairs.

(A pair of polynomials (N, D) is said to be coprime if N and D do not have common roots.) For example, let $P(s) = \frac{(s-1)}{s(s^2+s+1)}$; in this case we can choose $N_P(s) = (s - 1)$, $D_P(s) = s(s^2 + s + 1)$. Note that $N_P(s) = -(s - 1)$ and $D_P(s) = -s(s^2 + s + 1)$ would also be a feasible choice, but there is no other possibility, because N_P and D_P are not allowed to have common roots. Now it is clear that the feedback system is stable if and only if the roots of

$$\chi(s) := D_P(s)D_C(s) + N_P(s)N_C(s) = 0 \qquad (4.9)$$

are in \mathbb{C}_-. The polynomial $\chi(s)$ is called the *characteristic polynomial*, and its roots are the closed-loop system poles. A polynomial is said to be stable (or *Hurwitz stable*) if its roots are in \mathbb{C}_-. Once a controller is specified for a given plant, the closed-loop system stability can easily be determined by constructing $\chi(s)$, and by computing its roots.

Example 4.1 Consider the plant $P(s) = \frac{(s-1)}{s\,(s^2+s+1)}$, with a controller $C(s) = \frac{0.06\,(s-0.75)}{(s+0.5)}$. The roots of the characteristic polynomial

$$\chi(s) = s^4 + 1.5s^3 + 1.56s^2 + 0.395s + 0.045$$

are $-0.6 \pm j0.9$ and $-0.15 \pm j0.13$. So, the feedback system is stable. The roots of $\chi(s)$ are computed in MATLAB. Feedback system stability analysis can be done likewise by using any computer program that solves the roots of a polynomial.

Since $P(s)$ and $C(s)$ are fixed in the above analysis, the coefficients of $\chi(s)$ are known and fixed. There are some cases where plant parameters are not known precisely. For example,

$$P(s) = \frac{(\alpha - s)}{s(s + 2)}$$

where α is known to be in the interval $(0 , \alpha_{max})$, yet its exact value is unknown. The upper bound α_{max} represents the largest uncertainty level for this parameter. Let the controller be in the form

$$C(s) = \frac{1}{(s + \beta)}.$$

The parameter β is to be adjusted so that the feedback system is stable for all values of $\alpha \in (0 , \alpha_{max})$, i.e., the roots of

$$\chi(s) = s(s + 2)(s + \beta) + (\alpha - s) = s^3 + (2 + \beta)s^2 + (2\beta - 1)s + \alpha$$

are in \mathbb{C}_- for all $\alpha \in (0 , \alpha_{max})$. This way, closed-loop stability is guaranteed for the uncertain plant. For each fixed pair of parameters (α, β), the roots of $\chi(s)$ can be computed numerically. Hence, in the two-dimensional parameter space, the stability region can be determined by checking the roots of $\chi(s)$ at each (α, β) pair. For each fixed β, the largest feasible $\alpha > 0$ can easily be determined by a line search. Figure 4.3 shows α_{max} for each β. For $\beta \leq 0.5$, the feedback system is unstable. When $\beta > 0.5$ the largest allowable α increases non-linearly with β. The exact relationship between α_{max} and β will be determined from the Routh-Hurwitz stability test in the next section.

The numerical approach illustrated above can still be used even if there are more than two parameters involved in the characteristic polynomial. However, in that case, computational complexity (the number of grid points to be taken in the parameter space for checking the roots) grows exponentially with the number of parameters.

Figure 4.3: β versus α_{\max}

4.4 Routh-Hurwitz Stability Test

The Routh-Hurwitz test is a direct procedure for checking stability of a polynomial without computing its roots. Consider a polynomial of degree n with real coefficients a_0, \ldots, a_n:

$$\chi(s) = a_0 s^n + a_1 s^{n-1} + \cdots + a_n, \quad \text{where} \quad a_0 \neq 0.$$

The polynomial $\chi(s)$ is stable if and only if the number of sign changes in the first column of the Routh table is zero. The Routh table is constructed as follows:

$$
\begin{array}{ccccc}
a_0 & a_2 & a_4 & a_6 & \cdots \\
a_1 & a_3 & a_5 & a_7 & \cdots \\
R_{3,1} & R_{3,2} & R_{3,3} & R_{3,4} & \cdots \\
R_{4,1} & R_{4,2} & R_{4,3} & R_{4,4} & \cdots \\
\vdots & \vdots & \vdots & \vdots & \\
R_{n+1,1} & & & &
\end{array}
$$

The first two rows of this table are determined directly from the coefficients of the polynomial. For $k \geq 3$ the kth row is constructed from the $k - 1$st and $k - 2$nd rows according to the formula

$$R_{k,\ell} = \frac{R_{k-2,\ell+1} R_{k-1,1} - R_{k-1,\ell+1} R_{k-2,1}}{R_{k-1,1}} \qquad k \geq 3, \ \ell \geq 1.$$

We should set $a_\ell = 0$ for $\ell > n$ for constructing the 3rd, 4th, and remaining rows. If n is even, the length of the first row is $\ell_1 = \frac{n}{2} + 1$, and the length of the second row is $\ell_2 = \ell_1 - 1$; if n is odd then $\ell_1 = \frac{n+1}{2}$, and $\ell_2 = \ell_1$. Then, for $k \geq 3$ the length of kth row, ℓ_k, can be determined as $\ell_k = \ell_{k-2} - 1$. Therefore, the Routh table has a block upper triangular form. Another important point to note is that if $R_{k-1,1}$ is zero, then the kth row cannot be determined from the above formula because of the division. A zero element in the first column of the Routh table indicates existence of a root in $\overline{\mathbb{C}}_+$, so the polynomial is unstable in that case. Suppose that there is no zero element in the first column. Then the number of roots of $\chi(s)$ in \mathbb{C}_+ is equal to the number of sign changes in the first column of the Routh table.

Example 4.2 Consider the polynomial $\chi(s) = s^3 + (2 + \beta)s^2 + (2\beta - 1)s + \alpha$, for which the Routh table is

$$\begin{array}{cc} 1 & 2\beta - 1 \\ 2 + \beta & \alpha \\ R_{3,1} & 0 \\ R_{4,1} & \end{array}$$

where $R_{4,1} = \alpha$, and $R_{3,1} = (2\beta - 1) - \alpha(2 + \beta)^{-1}$. So, for stability of $\chi(s)$ we need $\alpha > 0$, $\beta > -2$, and $(2\beta - 1)(2 + \beta) > \alpha > 0$, which imply $\beta > 0.5$. These inequalities are in agreement with Figure 4.3. It is now clear that the exact relationship between β and α_{\max} is

$$\alpha_{\max} = 2\beta^2 + 3\beta - 2 \quad \text{for} \quad \beta > 0.5.$$

Exercise: The unstable plant

$$P(s) = \frac{1}{(s-1)(s^2 + 2s + 2)}$$

is to be compensated by the controller

$$C(s) = \frac{K(s+2)}{(s+3)}.$$

The gain K will be adjusted for closed-loop system stability. By using the Routh-Hurwitz test, show that the feedback system is stable if and only if $3 < K < 14$ and $(68 - K - K^2) > 0$. So the gain K should be selected from the range $3 < K < \sqrt{68.25} - 0.5 = 7.76$.

4.5 Stability Robustness: Parametric Uncertainty

The Routh-Hurwitz stability test determines stability of a characteristic polynomial $\chi(s)$ with *fixed* coefficients. If there are only a few uncertain parameters (due to plant uncertainty) or free parameters (of controller) in the coefficients of $\chi(s)$, then it is still possible to use the Routh-Hurwitz test to determine the set of all admissible parameters for stability. This point was illustrated with the above examples. If the number of variable parameters is large, then the analysis is cumbersome. In this section, we will see simple robust stability tests for plants with uncertain coefficients in the transfer function.

4.5.1 Uncertain Parameters in the Plant

Recall that $P(s)$ is a mathematical model of the physical system. Depending on the confidence level with this model, each coefficient of $N_P(s)$ and $D_P(s)$ can be assumed to lie in an interval of the real line.

For example, taking only one mode of a flexible beam, its transfer function (from force input to acceleration output) can be written as

$$P(s) = \frac{K_p}{s^2 + 2\zeta\omega_o s + \omega_o^2} \; .$$

Suppose $K_p \in [0.8 \; , \; 1.3]$, $\zeta \in [0.2 \; , \; 0.3]$, and $\omega_o \in [10 \; , \; 12]$. Then, $N_P(s) = q_0$, where $q_0 \in [0.8 \; , \; 1.3]$, and $D_P(s) = r_0 s^2 + r_1 s + r_2$, where $r_0 \in [1 \; , \; 1]$, $r_1 \in [4 \; , \; 7.2]$, $r_2 \in [100 \; , \; 144]$.

Generalizing this representation, we assume that $P = \frac{N_P}{D_P}$ where

$$
\begin{aligned}
N_P(s) &= q_0 s^m + q_1 s^{m-1} + \cdots + q_m \; , \quad q_k \in [q_k^- \; , \; q_k^+] \; , \; k = 0, ..., m; \\
D_P(s) &= r_0 s^n + r_1 s^{n-1} + \cdots + r_n \quad r_\ell \in [r_\ell^- \; , \; r_\ell^+] \; , \; \ell = 0, ..., n.
\end{aligned}
$$

Since $P(s)$ is proper, $n \geq m$. The set of all possible plant transfer functions (determined from the set of all possible coefficients of $N_P(s)$ and $D_P(s)$) is denoted by $\mathcal{P}_{q,r}$. More precisely,

$$\mathcal{P}_{q,r} = \left\{ \frac{N_P(s)}{D_P(s)} = \frac{q_0 s^m + q_1 s^{m-1} + \cdots + q_m}{r_0 s^n + r_1 s^{n-1} + \cdots + r_n} \; : \; \begin{array}{l} q_k \in [q_k^- \; , \; q_k^+] \\ r_\ell \in [r_\ell^- \; , \; r_\ell^+] \end{array} \right\}$$

where $r_0^-, r_0^+, \ldots, r_m^-, r_m^+, q_0^-, q_0^+, \ldots, q_n^-, q_n^+$ are given upper and lower bounds of the plant parameters. In the literature, the class of uncertain plants of the form $\mathcal{P}_{q,r}$ is called *interval plants*. Total number of uncertain parameters in this description is $(n+m+2)$. In the parameter space $\mathbb{R}^{(n+m+2)}$, the set of all possible parameters is a "multi-dimensional box," for example, when $n + m + 2 = 2$ the set is a rectangle, when $n+m+2 = 3$ the set is a three dimensional box, and for $(n+m+2) \geq 4$ the set is a *polytope*.

The feedback system formed by a fixed controller $C = \frac{N_C}{D_C}$ and an uncertain plant $P \in \mathcal{P}_{q,r}$ is said to be *robustly stable* if all the roots of

$$\chi(s) = D_C(s)D_P(s) + N_C(s)N_P(s)$$

are in \mathbb{C}_- for all $P = \frac{N_P}{D_P} \in \mathcal{P}_{q,r}$.

For each q_k, $k = 0, \ldots, m$, we assume that it can take any value within the given interval, independent of the values of the other coefficients. Same assumption is made for r_ℓ, $\ell = 0, \ldots, n$. For example, if two parameters are related by, say, $r_0 = 3x - 1$ and $r_1 = 1 + x - 2x^2$, where $x \in [0, 1]$, then a conservative assumption would be $r_0 \in [-1, 2]$ and $r_1 \in [0, 1.125]$. In the (r_0, r_1) parameter space, the set R_{r_0, r_1} is a rectangle that includes the line L_{r_0, r_1}:

$$R_{r_0, r_1} = \{(r_0, r_1) : r_0 \in [-1, 2], \ r_1 \in [0, 1.125]\}$$

$$L_{r_0, r_1} = \{(r_0, r_1) : r_0 = 3x - 1, \ r_1 = 1 + x - 2x^2, \ x \in [0, 1]\}.$$

If the closed-loop system is stable for all values of (r_0, r_1) in the set R_{r_0, r_1}, then it is stable for all values of (r_0, r_1) in L_{r_0, r_1}. However, the converse is not true, i.e., there may be a point in R_{r_0, r_1} for which the system is not stable, while the system might be stable for all points in L_{r_0, r_1}. This is the conservatism in transforming a dependent parameter uncertainty to an independent parameter uncertainty.

4.5.2 Kharitanov's Test for Robust Stability

Consider a feedback system formed by a fixed controller $C(s) = \frac{N_C(s)}{D_C(s)}$ and an uncertain plant $P(s) = \frac{N_P(s)}{D_P(s)} \in \mathcal{P}_{q,r}$. To test robust stability, first construct the characteristic polynomial

$$\chi(s) = D_C(s)D_P(s) + N_C(s)N_P(s)$$

with uncertain coefficients. Note that from the upper and lower bounds of the coefficients of $D_P(s)$ and $N_P(s)$ we can determine (albeit in a conservative fashion) upper and lower bounds of the coefficients of $\chi(s)$.

Example 4.3 Let $N_C(s) = (s+2)$, $D_C(s) = (s^2 + 2s + 2)$ and $D_P(s) = r_0 s^3 + r_1 s^2 + r_2 s + r_3$, $N_P(s) = q_0$, with $r_0 \in [1, 1.1]$, $r_1 \in [4, 4.2]$, $r_2 \in [6, 8]$, $r_3 \in [10, 20]$, and $q_0 \in [3, 5]$. Then $\chi(s)$ is in the form

$$\chi(s) = (s^2 + 2s + 2)(r_0 s^3 + r_1 s^2 + r_2 s + r_3) + (s + 2)(q_0)$$

$$
\begin{aligned}
&= r_0 s^5 + (r_1 + 2r_0)s^4 + (r_2 + 2r_1 + 2r_0)s^3 \\
&\quad + (r_3 + 2r_2 + 2r_1)s^2 + (2r_3 + 2r_2 + q_0)s + (2r_3 + 2q_0) \\
&= a_0 s^5 + a_1 s^4 + a_2 s^3 + a_3 s^2 + a_4 s + a_5
\end{aligned}
$$

where $a_0 \in [1 , 1.1]$, $a_1 \in [6 , 6.4]$, $a_2 \in [16 , 18.6]$, $a_3 \in [30 , 44.4]$, $a_4 \in [35 , 61]$, $a_5 \in [26 , 50]$. We assume that the parameter vector $[a_0, \ldots, a_5]$ can take any values in the subset of \mathbb{R}^6 determined by the above intervals for each component. In other words, the coefficients vary independent of each other. However, there are only five truly free parameters (r_0, \ldots, r_3, q_0), which means that there is a dependence between parameter variations. If robust stability can be shown for all possible values of the parameters a_k in the above intervals, then robust stability of the closed-loop system can be concluded; but the converse is not true. In that sense, by converting plant (and/or controller) parameter uncertainty into an uncertainty in the coefficients of the characteristic polynomial, some conservatism is introduced.

Now consider a typical characteristic polynomial

$$
\chi(s) = a_0 s^N + a_1 s^2 + \ldots + a_N
$$

where each coefficient a_k can take any value in a given interval $[a_k^- , a_k^+]$, $k = 0, 1, \ldots, N$, independent of the values of a_j, $j \neq k$. The set of all possible characteristic equations is defined by the upper and lower bounds of each coefficient. It will be denoted by \mathcal{X}_a, i.e.

$$
\mathcal{X}_a := \{ a_0 s^N + \ldots + a_N \ : \ a_k \in [a_k^- , a_k^+] , k = 0, 1, \ldots, N \}.
$$

Theorem 4.3 Kharitanov's Theorem. . *All polynomials in \mathcal{X}_a are stable if and only if the following four polynomials* $a_1(s), \ldots, a_4(s)$ *are stable:*

$$
\begin{aligned}
a_1(s) &= a_N^- + a_{N-1}^- s + a_{N-2}^+ s^2 + a_{N-3}^+ s^3 + a_{N-4}^- s^4 + a_{N-5}^- s^5 + \cdots \\
a_2(s) &= a_N^+ + a_{N-1}^+ s + a_{N-2}^- s^2 + a_{N-3}^- s^3 + a_{N-4}^+ s^4 + a_{N-5}^+ s^5 + \cdots
\end{aligned}
$$

$$\mathsf{a_3}(s) \;=\; a_N^- + a_{N-1}^+ s + a_{N-2}^+ s^2 + a_{N-3}^- s^3 + a_{N-4}^- s^4 + a_{N-5}^+ s^5 + \cdots$$

$$\mathsf{a_4}(s) \;=\; a_N^+ + a_{N-1}^- s + a_{N-2}^- s^2 + a_{N-3}^+ s^3 + a_{N-4}^+ s^4 + a_{N-5}^- s^5 + \cdots$$

In the literature, the polynomials $\mathsf{a_1}(s), \ldots, \mathsf{a_4}(s)$ are called *Kharit-anov polynomials*. By virtue of Kharitanov's theorem, robust stability can be checked by applying the Routh-Hurwitz stability test on four polynomials. Considering the complexity of the parameter space, this is a great simplification. For easily accessible proofs of Kharitanov's theorem see [5, pp. 70-78] and [9, pp. 224-229].

4.5.3 Extensions of Kharitanov's Theorem

Kharitanov's theorem gives necessary and sufficient conditions for robust stability of the set \mathcal{X}_a. On the other hand, recall that for a fixed controller and uncertain plant in the set $\mathcal{P}_{\mathsf{q,r}}$ the characteristic polynomial is in the form

$$\chi(s) \;=\; D_C(s)\; (r_0 s^n + r_1 s^{n-1} + \ldots + r_n)$$
$$+ \, N_C(s)\; (q_0 s^m + q_1 s^{m-1} + \ldots + q_m)$$

where $r_k \in [r_k^-, r_k^+]$, $k = 0, \ldots, n$, and $q_\ell \in [q_\ell^-, q_\ell^+]$, $\ell = 0, \ldots, m$. Let us denote the set of all possible characteristic polynomials corresponding to this uncertainty structure by $\mathcal{X}_{\mathsf{q,r}}$. As seen in the previous section, it is possible to define a larger set of all possible characteristic polynomials, denoted by \mathcal{X}_a, and apply Kharitanov's theorem to test robust stability. However, since $\mathcal{X}_{\mathsf{q,r}}$ is a proper subset of \mathcal{X}_a, Kharitanov's result becomes a conservative test. In other words, if four Kharitanov polynomials (determined from the larger uncertainty set \mathcal{X}_a) are stable, then all polynomials in $\mathcal{X}_{\mathsf{q,r}}$ are stable; but the converse is not true, i.e., one of the Kharitanov polynomials may be unstable, while all polynomials in $\mathcal{X}_{\mathsf{q,r}}$ are stable. There exists a non-conservative test for robust stability of the polynomials in $\mathcal{X}_{\mathsf{q,r}}$; it is given by the result stated below, called the 32 edge theorem [5], or generalized Kharitanov's theorem [9].

First define $N_1(s), \ldots, N_4(s)$, four Kharitanov polynomials corresponding to uncertain polynomial $N_P(s) = q_0 s^m + \ldots + q_m$, and similarly define $D_1(s), \ldots, D_4(s)$, four Kharitanov polynomials corresponding to uncertain polynomial $D_P(s) = r_0 s^n + \ldots + r_n$. Then, for all possible combinations of $i_1 \in \{1, 2, 3, 4\}$, and $(i_2, i_3) \in \{(1,3), (1,4), (2,3), (2,4)\}$ define 16 polynomials, which depend on a parameter λ,

$$e_{1-16}(s, \lambda) = N_{i_1}(s)N_C(s) + (\lambda D_{i_2}(s) + (1 - \lambda)D_{i_3}(s))D_C(s).$$

Similarly, for all possible combinations of $i_3 \in \{1, 2, 3, 4\}$, and $(i_1, i_2) \in \{(1,3), (1,4), (2,3), (2,4)\}$ define the next set of 16 λ dependent polynomials,

$$e_{17-32}(s, \lambda) = D_{i_3}(s)D_C(s) + (\lambda N_{i_1}(s) + (1 - \lambda)N_{i_2}(s))N_C(s).$$

Theorem 4.4 [9] *Assume that all the polynomials in $\mathcal{X}_{q,r}$ have the same degree. Then, all polynomials in $\mathcal{X}_{q,r}$ are stable if and only if $e_1(s, \lambda), \ldots, e_{32}(s, \lambda)$ are stable for all $\lambda \in [0 , 1]$.*

This result gives a necessary and sufficient condition for robust stability of polynomials in $\mathcal{X}_{q,r}$. The test is more complicated than Kharitanov's robust stability test; it involves checking stability of 32 polynomials for all values of $\lambda \in [0 , 1]$. For each $e_k(s, \lambda)$ it is easy to construct the Routh table in terms of λ and test stability of e_k for all $\lambda \in [0 , 1]$. This is a numerically feasible test. However, since there are infinitely many possibilities for λ, technically speaking one needs to check stability of infinitely many polynomials. For the special case where the controller is fixed as a first order transfer function

$$C(s) = \frac{K_c(s - z)}{(s - p)} \text{ , where } K_c, \ z, \ p, \text{ are fixed and } z \neq p,$$

the test can be reduced to checking stability of 16 polynomials only.

Using the above notation, let $N_1(s), \ldots, N_4(s)$ and $D_1(s), \ldots, D_4(s)$ be the Kharitanov polynomials for $N_P(s) = (q_0 s^m + \ldots + q_m)$ and

$D_P(s) = (r_0 s^n + \ldots + r_n)$, respectively. Define $\mathcal{P}_{q,r}$ as the set of all plants and assume that $0 \neq [r_0^-, r_0^+]$, i.e., the degree of $D_P(s)$ is fixed.

Theorem 4.5 [5]. *The closed-loop system formed by the plant P and the controller $\frac{N_C}{D_C}$, where $N_C(s) = K_c(s - z)$ and $D_C(s) = (s - p)$, is stable for all $P \in \mathcal{P}_{q,r}$ if and only if the following 16 polynomials are stable:*

$$N_C(s)N_{i_1}(s) + D_C(s)D_{i_2}(s)$$

where $i_1 \in \{1, 2, 3, 4\}$ and $i_2 \in \{1, 2, 3, 4\}$.

This result is called the 16 plant theorem, and it remains valid for slightly more general cases in which the controller N_C/D_C is

$$N_C(s) = K_c(s - z)U_N(s)R_N(s) \tag{4.10}$$
$$D_C(s) = s^\ell(s - p)U_D(s)R_D(s) \tag{4.11}$$

where K_c, z, p are real numbers, $\ell \geq 0$ is an integer, $U_N(s)$ and $U_D(s)$ are anti-stable polynomials (i.e., all roots in \mathbb{C}_+), and $R_N(s)$ and $R_D(s)$ are in the form $R(s^2)$, where $R(s)$ is an arbitrary polynomial, [9]. When the controller is restricted to this special structure, 32 edge theorem reduces to checking stability of the closed loop systems formed by the controller N_C/D_C and 16 plants $P(s) = N_{i_1}(s)/D_{i_2}(s)$, for $i_1 \in \{1, 2, 3, 4\}$, and $i_2 \in \{1, 2, 3, 4\}$.

For the details and proofs of 32 edge theorem and 16 plant theorem see [5, pp. 153-195], and [9, pp. 300-334].

4.6 Exercise Problems

1. Given a characteristic polynomial $\chi(s) = a_0 s^3 + a_1 s^2 + a_2 s + a_3$ with coefficients in the intervals $0.1 < a_3 < 0.5$; $1 < a_2 < 2$;

$1 < a_1 < 3$; $1 < a_0 < 4$. Using Kharitanov's test, show that we do not have robust stability. Now suppose a_1 and a_0 satisfy

$$a_1 = 1 + 2x^2 \quad a_0 = 1 + 3x^2 \quad \text{where} \quad -1 < x < 1 .$$

Do we have robust stability? Hint: Use the Routh Hurwitz test here; Kharitanov's test does not give a conclusive answer.

2. Consider the standard feedback control system with an interval plant $P \in \mathcal{P}_{q,r}$,

$$\mathcal{P}_{q,r} = \left\{ P(s) = \frac{q_0 s^3 + q_1 s^2 + q_2 s + q_3}{r_0 s^4 + r_1 s^3 + r_2 s^2 + r_3 s + r_4} \right\}$$

where $q_0 \in [0.95 , 1.25]$, $q_1 \in [-4.0 , -3.6]$, $q_2 \in [3.0 , 3.5]$, $q_3 \in [0.6 , q]$, $r_0 \in [0.9 , 1.1]$, $r_1 \in [11.5 , 12.5]$, $r_2 \in [20 , 24]$, $r_3 \in [16 , 24]$, $r_4 \in [-0.1 , 0.1]$. By using the 16 plant theorem find the maximum value of q such that there exists a robustly stabilizing controller of the form

$$C(s) = \frac{K (s + 1)}{s}$$

for the family of plants $\mathcal{P}_{q,r}$. Determine the corresponding value of K.

Chapter 5

Root Locus

Recall that the roots of the characteristic polynomial

$$\chi(s) = D_P(s)D_C(s) + N_P(s)N_C(s)$$

are the poles of the feedback system formed by the controller $C = N_C/D_C$ and the plant $P = N_P/D_P$. In Chapter 3 we saw that in order to achieve a certain type of performance objectives, the dominant closed-loop poles must be placed in a specified region of the complex plane. Once the pole-zero structure of $G(s) = P(s)C(s)$ is fixed, the gain of the controller can be adjusted to see whether the design specifications are met with this structural choice of $G(s)$. In the previous chapter we also saw that robust stability can be tested by checking stability of a family of characteristic polynomials depending on a parameter (e.g. λ of the 32-edge theorem). In these examples, the characteristic polynomial is an affine function of a parameter. The root locus shows the closed-loop system poles as this parameter varies.

Numerical tools, e.g., MATLAB, can be used to construct the root locus with respect to a parameter that appears nonlinearly in the char-

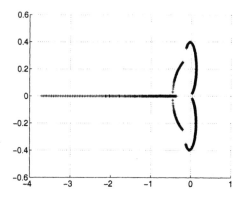

Figure 5.1: Root locus as a function of $h \in [1, 100]$.

acteristic polynomial. For example, consider the plant

$$P(s) = \frac{((sh)^2/12 - sh/2 + 1)}{s\,((sh)^2/12 + sh/2 + 1)} \tag{5.1}$$

as an approximation of a system with a time delay and an integrator. Let the controller for the plant $P(s)$ be

$$C(s) = 0.3 + \frac{0.1}{s}.$$

Then the characteristic polynomial

$$\chi(s) = (\frac{h^2}{12}s^2 + 1)(s^2 + 0.3s + 0.1) + \frac{h}{2}\,s\,(s^2 - 0.3s - 0.1)$$

is a nonlinear function of h, due to h^2 terms. Figure 5.1 shows the closed-loop system pole locations as h varies from a lower bound $h_{\min} = 1$ to an upper bound $h_{\max} = 100$. The figure is obtained by computing the roots of $\chi(s)$ for a set of values of $h \in [h_{\min}, h_{\max}]$.

As mentioned above, the root locus primarily deals with finding the roots of a characteristic polynomial that is an affine function of a single parameter, say K,

$$\chi(s) = D(s) + KN(s) \tag{5.2}$$

where $D(s)$ and $N(s)$ are fixed monic polynomials (i.e., coefficient of the highest power is normalized to 1). In particular, 32-edge polynomials are in this form. For example, recall that

$$
\begin{aligned}
e_3(s, \lambda) &= N_1(s)N_C(s) + (\lambda D_2(s) + (1 - \lambda)D_3(s))D_C(s) \\
&= (N_1(s)N_C(s) + D_3(s)D_C(s)) + \lambda(D_2(s) - D_3(s))D_C(s).
\end{aligned}
$$

By defining $K = \frac{\lambda}{1-\lambda}$ (i.e. $\lambda = \frac{K}{K+1}$), $N = (N_1 N_C + D_3 D_C)$ and $D = (N_1 N_C + D_2 D_C)$, it can be shown that the roots of $e_3(s, \lambda)$ over the range of $\lambda \in [0, 1]$ are the roots of $\chi(s)$ defined in (5.2) over the range of $K \in [0, +\infty]$. If N and D are not monic, the highest coefficient of D can be factored out of the equation and the ratio of the highest coefficient of N to that of D can be absorbed into K.

As another example, consider a fixed plant $P = N_P/D_P$ and a PI controller with fixed proportional gain K_p and variable integral gain K_i

$$
C(s) = K_p + \frac{K_i}{s}.
$$

The characteristic equation is

$$
\chi(s) = s(D_P(s) + K_p N_P(s)) + K_i N_P(s) ,
$$

which is in the form (5.2) with $K = K_i$, $D(s) = s\ (D_P(s) + K_p N_P(s))$ and $N(s) = N_P(s)$.

The most common example of (5.2) is the variable controller gain case: when the controller and plant are expressed in the pole-zero form as

$$
\begin{aligned}
P(s) &= K_P \frac{(s - z_{i_1}) \ldots (s - z_{i_m})}{(s - p_{i_1}) \ldots (s - p_{i_n})} \\
C(s) &= K_C \frac{(s - z_{j_1}) \ldots (s - z_{j_m})}{(s - p_{j_1}) \ldots (s - p_{j_n})}
\end{aligned}
$$

the characteristic equation is as (5.2) with $K = K_P K_C$, and

$$
\begin{aligned}
D(s) &= (s - z_{i_1}) \ldots (s - z_{i_m})(s - z_{j_1}) \ldots (s - z_{j_m}) \\
N(s) &= (s - p_{i_1}) \ldots (s - p_{i_n})(s - p_{j_1}) \ldots (s - p_{j_n})
\end{aligned}
$$

To simplify the notation set $m := i_m + j_m$, and $n := i_n + j_n$, and enumerate poles and zeros of $G(s) = P(s)C(s)$ in such a way that

$$G(s) = K \frac{(s - z_1) \ldots (s - z_m)}{(s - p_1) \ldots (s - p_n)} = K \frac{N(s)}{D(s)},$$

then assuming $K \neq 0$ the characteristic equation (5.2) is equivalent to

$$1 + G(s) = 0 \iff 1 + KG_1(s) = 0 \tag{5.3}$$

where $G_1(s) = N(s)/D(s)$, which is equal to $G(s)$ evaluated at $K = 1$. The purpose of this chapter is to examine how closed-loop system poles (roots of the characteristic equation that is either in the form (5.2), or (5.3)) change as K varies from 0 to $+\infty$, or from 0 to $-\infty$.

5.1 Root Locus Rules

The usual root locus (abbreviated as RL) shows the locations of the closed-loop system poles as K varies from 0 to $+\infty$. The roots of $D(s)$, p_1, \cdots, p_n, are the poles of the open-loop system $G(s)$, and the roots of $N(s)$, z_1, \cdots, z_m, are the zeros of $G(s)$. Since $P(s)$ and $C(s)$ are proper, $G(s)$ is proper and hence $n \geq m$. So the degree of the polynomial $\chi(s)$ is n and it has exactly n roots.

Let the closed-loop system poles, i.e., roots of $\chi(s)$, be denoted by $r_1(K), \cdots, r_n(K)$. Note that these are functions of K; whenever the dependence on K is clear, they are simply written as r_1, \ldots, r_n. The points in \mathbb{C} that satisfy (5.3) for some $K > 0$ are on the RL. Clearly, a point $r \in \mathbb{C}$ is on the RL if and only if

$$K = -\frac{1}{G_1(r)} \quad . \tag{5.4}$$

The condition (5.4) can be separated into two parts:

$$|K| = \frac{1}{|G_1(r)|} \tag{5.5}$$

$$\angle K = 0° \quad = \quad -(2\ell + 1)180° - \angle G_1(r), \quad \ell = 0, \pm 1, \pm 2, \ldots. \quad (5.6)$$

The phase rule (5.6) determines the points in \mathbb{C} that are on the RL. The magnitude rule (5.5) determines the gain $K > 0$ for which the RL is at a given point r. By using the definition of $G_1(s)$, (5.6) can be rewritten as

$$(2\ell + 1)180° = \sum_{i=1}^{n} \angle(r - p_i) - \sum_{j=1}^{m} \angle(r - z_j). \quad (5.7)$$

Similarly, (5.5) is equivalent to

$$K = \frac{\prod_{i=1}^{n} |r - p_i|}{\prod_{j=1}^{m} |r - z_j|}. \quad (5.8)$$

5.1.1 Root Locus Construction

There are several software packages available for generating the root locus automatically for a given $G_1 = N/D$. In particular, the related MATLAB commands are rlocus and rlocfind. In many cases, approximate root locus can be drawn by hand using the rules given below. These rules are determined from the basic definitions (5.2), (5.5) and (5.6).

1. The root locus has n branches: $r_1(K), \cdots, r_n(K)$.

2. Each branch starts ($K \cong 0$) at a pole p_i and ends (as $K \to \infty$) at a zero z_j, or converges to an asymptote, $Re^{j\alpha_\ell}$, where $R \to \infty$ and α_ℓ is determined from the formula

$$(n - m)\alpha_\ell = (2\ell + 1)180°, \quad \ell = 0, \cdots, (n - m - 1).$$

3. There are $(n - m)$ asymptotes with angles α_ℓ. The center of the asymptotes (i.e., their intersection point on the real axis) is

$$\sigma_a = \frac{\left(\sum_{i=1}^{n} p_i\right) - \left(\sum_{j=1}^{m} z_j\right)}{n - m}.$$

4. A point $x \in \mathbb{R}$ is on the root locus if and only if the total number of poles p_i's and zeros z_j's to the right of x (i.e., total number of p_i's with $\text{Re}(p_i) > x$ plus total number of z_j's with $\text{Re}(z_j) > x$) is odd. Since $G_1(s)$ is a rational function with real coefficients, poles and zeros appear in complex conjugates, so when counting the number of poles and zeros to the right of a point $x \in \mathbb{R}$ we just need to consider the poles and zeros on the real axis.

5. The values of K for which the root locus crosses the imaginary axis can be determined from the Routh-Hurwitz stability test. Alternatively, we can set $s = j\omega$ in (5.2) and solve for real ω and K satisfying

$$D(j\omega) + KN(j\omega) = 0.$$

Note that there are two equations here, one for the real part and one for the imaginary part.

6. The break points (intersection of two branches on the real axis) are feasible solutions (satisfying rule #4) of

$$\frac{d}{ds}G_1(s) = 0. \tag{5.9}$$

7. Angles of departure ($K \cong 0$) from a complex pole, or arrival ($K \to +\infty$) to a complex zero, can be determined from the phase rule. See example below.

Let us now follow the above rules step by step to construct the root locus for

$$G_1(s) = \frac{(s + 3)}{(s - 1)(s + 5)(s + 4 + j2)(s + 4 - j2)}.$$

First, enumerate the poles and zeros as $p_1 = -4 + j2$, $p_2 = -4 - j2$, $p_3 = -5$, $p_4 = 1$, $z_1 = -3$. So, $n = 4$ and $m = 1$.

1. The root locus has four branches.

2. Three branches converge to the asymptotes whose angles are $60°$, $180°$ and $-60°$, and one branch converges to $z_1 = -3$.

3. Center of the asymptotes is $\sigma = (-12 + 3)/3 = -3$.

4. The intervals $(-\infty, -5]$ and $[-3, 1]$ are on the root locus.

5. The imaginary axis crossings are the feasible roots of

$$(\omega^4 - j12\omega^3 - 47\omega^2 + j40\omega - 100) + K(j\omega + 3) = 0 \qquad (5.10)$$

for real ω and K. Real and imaginary parts of (5.10) are

$$\omega^4 - 47\omega^2 - 100 + 3K = 0$$
$$j\omega(-12\omega^2 + 40 + K) = 0.$$

They lead to two feasible pair of solutions $(K = \frac{100}{3}, \omega = 0)$ and $(K = 215.83, \omega = \pm4.62)$.

6. Break points are the feasible solutions of

$$3s^4 + 36s^3 + 155s^2 + 282s + 220 = 0.$$

Since the roots of the above polynomial are $-4.55 \pm j1.11$ and $-1.45 \pm j1.11$, there is no solution on the real axis, hence no break points.

7. To determine the angle of departure from the complex pole $p_1 = -4 + j2$ let Δ represent a point on the root locus near the complex pole p_1, and define v_i, $i = 1, \ldots, 5$, to be the vectors drawn from p_i, for $i = 1, \ldots, 4$, and from z_1 for $i = 5$, as shown in Figure 5.2. Let $\theta_1, \ldots, \theta_5$ be the angles of v_1, \ldots, v_5. The phase rule implies

$$(\theta_1 + \theta_2 + \theta_3 + \theta_4) - \theta_5 = \pm180°. \qquad (5.11)$$

As Δ approaches to p_1, θ_1 becomes the angle of departure and the other θ_i's can be approximated by the angles of the vectors drawn

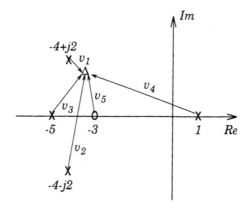

Figure 5.2: Angle of departure from $-4 + j2$.

from the other poles, and from the zero, to the pole p_1. Thus θ_1 can be solved from (5.11) where $\theta_2 \approx 90°$, $\theta_3 \approx \tan^{-1}(2)$, $\theta_4 \approx 180° - \tan^{-1}(\frac{2}{5})$, and $\theta_5 \approx 90° + \tan^{-1}(\frac{1}{2})$. That yields $\theta_1 \approx -15°$.

The exact root locus for this example is shown in Figure 5.3. From the results of item #5 above, and the shape of the root locus it is concluded that the feedback system is stable if

$$33.33 < K < 215.83$$

i.e., by simply adjusting the gain of the controller, the system can be made stable. In some situations we need to use a dynamic controller to satisfy all the design requirements.

5.1.2 Design Examples

Example 5.1 Consider the standard feedback system with a plant

$$P(s) = \frac{1}{0.72} \frac{1}{(s+1)(s+2)}$$

Figure 5.3: Root Locus for $G_1(s) = \dfrac{(s+3)}{(s-1)(s+5)(s+4+j2)(s+4-j2)}$.

and design a controller such that

- the feedback system is stable,

- PO $\leq 10\%$, $t_s \leq 4$ sec. and $e_{ss} = 0$ when $r(t) = \mathbb{U}(t)$

- e_{ss} is as small as possible when $r(t) = t\mathbb{U}(t)$.

It is clear that the second design goal (the part that says that e_{ss} should be zero for unit step reference input) cannot be achieved by a simple proportional controller. To satisfy this condition, the controller must have a pole at $s = 0$, i.e., it must have integral action. If we try an integral control of the form $C(s) = K_c/s$, with $K_c > 0$, then the root locus has three branches, the interval $[-1 , 0]$ is on the root locus; three asymptotes have angles $\{60°, 180°, -60°\}$ with a center at $\sigma_a = -1$; and there is only one break point at $-1 + \frac{1}{\sqrt{3}}$. See Figure 5.4. From the location of the break point, center, and angles of the asymptotes, it can be deduced that two branches (one starting at $p_1 = -1$, and the other one starting at $p_3 = 0$) always remain to the right of the point -1. On the other hand, the settling time condition implies that the real parts of the dominant closed-loop system poles must be less than

or equal to -1. So, a simple integral control does not do the job. Now try a PI controller of the form

$$C(s) = K_c \frac{(s - z_c)}{s} \qquad K_c > 0.$$

In this case, we can select $z_c = -1$ to cancel the pole at $p_1 = -1$ and the system effectively becomes a second-order system. The root locus for $G_1(s) = 1/s(s + 2)$ has two branches and two asymptotes, with center $\sigma_a = -1$ and angles $\{90°, -90°\}$; the break point is also at -1. The branches leave -2 and 0, and go toward each other, meet at -1, and tend to infinity along the line $\mathrm{Re}(s) = -1$. Indeed, the closed-loop system poles are

$$r_{1,2} = -1 \pm \sqrt{1 - K} \qquad \text{where } K = K_c/0.72 .$$

The steady state error, when $r(t)$ is unit ramp, is $2/K$. So K needs to be as large as possible to meet the third design condition. Clearly, $\mathrm{Re}(r_{1,2}) = -1$ for all $K \geq 1$, that satisfies the settling time requirement. The percent overshoot is less than 10% if ζ of the roots $r_{1,2}$ is greater than 0.6. A simple algebra shows that $\zeta = 1/\sqrt{K}$, hence the design conditions are met if $K = 1/0.36$, i.e. $K_c = 2$. Thus a PI controller that solves the design problem is

$$C(s) = 2 \frac{(s + 1)}{s} .$$

The controller cancels a stable pole (at $s = -1$) of the plant. If there is a slight uncertainty in this pole location, perfect cancelation will not occur and the system will be third-order with the third pole at $r_3 \cong -1$. Since the zero at $z_o = -1$ will approximately cancel the effect of this pole, the response of this system will be close to the response of a second-order system. However, we must be careful if the pole zero cancelations are near the imaginary axis because in this case small perturbations in pole location might lead to large variations in the feedback system response, as illustrated with the next example.

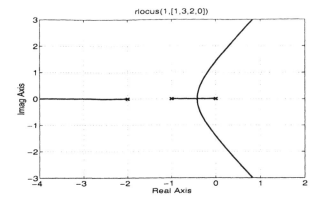

Figure 5.4: Root locus for Example 5.1.

Example 5.2 A flexible structure with lightly damped poles has transfer function in the form

$$P(s) = \frac{\omega_1^2}{s^2(s^2 + 2\zeta\omega_1 s + \omega_1^2)}.$$

By using the root locus, we can see that the controller

$$C(s) = K_c \frac{(s^2 + 2\zeta\omega_1 s + \omega_1^2)(s + 0.4)}{(s + r)^2(s + 4)}$$

stabilizes the feedback system for sufficiently large r and an appropriate choice of K_c. For example, let $\omega_1 = 2$, $\zeta = 0.1$ and $r = 10$. Then the root locus of $G_1(s) = P(s)C(s)/K$, where $K = K_c\omega_1^2$, is as shown in Figure 5.5. For $K = 600$ the closed-loop system poles are:

$$\{-10.78 \pm j2.57 \ , \ -0.94 \pm j1.61 \ , \ -0.2 \pm j1.99 \ , \ -0.56\}.$$

Since the poles $-0.2 \pm j1.99$ are canceled by a pair of zeros at the same point in the closed-loop system transfer function $T = G(1 + G)^{-1}$, the dominant poles are at -0.56 and $-0.94 \pm j1.61$ (they have relatively large negative real parts and the damping ratio is about 0.5).

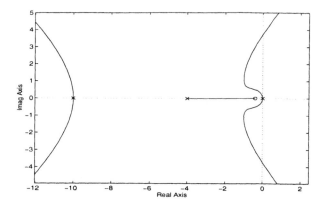

Figure 5.5: Root locus for Example 5.2 (a).

Now, suppose that this controller is fixed and the complex poles of the plant are slightly modified by taking $\zeta = 0.09$ and $\omega_1 = 2.2$. The root locus corresponding to this system is as shown in Figure 5.6. Since lightly damped complex poles are not perfectly canceled, there are two more branches near the imaginary axis. Moreover, for the same value of $K = 600$, the closed-loop system poles are

$$\{-10.78 \pm j2.57 \,, \ -1.21 \pm j1.86 \,, \ 0.05 \pm j1.93 \,, \ -0.51\}.$$

In this case, the feedback system is unstable.

Example 5.3 An approximate transfer function of a DC motor [37, pp. 141–143] is in the form

$$P_m(s) = \frac{K_m}{s\,(s + 1/\tau_m)} \,, \quad \tau_m > 0.$$

Note that if τ_m is large, then $P_m(s) \approx P_b(s)$, where

$$P_b(s) = \frac{K_b}{s^2}$$

is the transfer function of a rigid beam. In this example, the general class of plants $P_m(s)$ will be considered. Assuming that $p_m = \frac{-1}{\tau_m}$ and

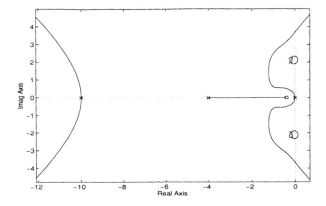

Figure 5.6: Root locus for Example 5.2 (b).

K_m are given, a first-order controller

$$C(s) = K_c \frac{(s - z_c)}{(s - p_c)} \tag{5.12}$$

will be designed. The aim is to place the closed-loop system poles far from the Im-axis. Since the order of $G_1(s) = P_m(s)C(s)/K_m K_c$ is three, the root locus has three branches. Suppose the desired closed loop poles are given as p_1, p_2 and p_3. Then, the pole placement problem amounts to finding $\{K_c, z_c, p_c\}$ such that the characteristic equation is

$$
\begin{aligned}
\chi(s) &= (s - p_1)(s - p_2)(s - p_3) \\
&= s^3 - (p_1 + p_2 + p_3)s^2 + (p_1 p_2 + p_1 p_3 + p_2 p_3)s - p_1 p_2 p_3.
\end{aligned}
$$

But the actual characteristic equation, in terms of the unknown controller parameters, is

$$
\begin{aligned}
\chi(s) &= s(s - p_m)(s - p_c) + K(s - z_c) \\
&= s^3 - (p_m + p_c)s^2 + (p_m p_c + K)s - K z_c
\end{aligned}
$$

where $K := K_m K_c$. Equating the coefficients of the desired $\chi(s)$ to the coefficients of the actual $\chi(s)$, three equations in three unknowns are obtained:

$$p_m + p_c = p_1 + p_2 + p_3$$

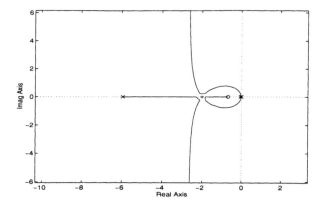

Figure 5.7: Root locus for Example 5.3 (a).

$$p_m p_c + K \;=\; p_1 p_2 + p_1 p_3 + p_2 p_3$$
$$K z_c \;=\; p_1 p_2 p_3$$

From the first equation p_c is determined, then K is obtained from the second equation, and finally z_c is computed from the third equation.

For different numerical values of p_m, p_1, p_2 and p_3 the shape of the root locus is different. Below are some examples, with the corresponding root loci shown in Figures 5.7–5.9.

(a) $p_m = -0.05$, $p_1 = p_2 = p_3 = -2 \Longrightarrow$

$$K = 11.70 \quad p_c = -5.95 \quad z_c = -0.68.$$

(b) $p_m = -0.5$, $p_1 = -1$, $p_2 = -2$, $p_3 = -3 \Longrightarrow$

$$K = 8.25 \quad p_c = -5.50 \quad z_c = -0.73.$$

(c) $p_m = -5$, $p_1 = -11$, $p_2 = -4 + j1$, $p_3 = -4 - j1 \Longrightarrow$

$$K = 35 \quad p_c = -14 \quad z_c = -5.343.$$

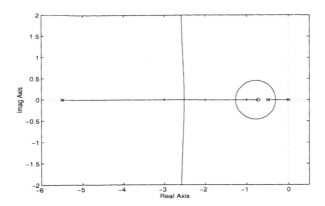

Figure 5.8: Root locus for Example 5.3 (b).

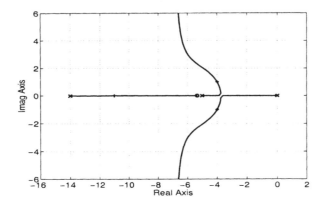

Figure 5.9: Root locus for Example 5.3 (c).

Example 5.4 The plant (5.1) with a fixed value of $h = 2$ will be controlled by using a first-order controller in the form (5.12). The open-loop transfer function is

$$P(s)C(s) = K_c \frac{(s^2 - 3s + 3)(s - z_c)}{s(s^2 + 3s + 3)(s - p_c)}$$

and the root locus has four branches. The first design requirement is to place the dominant poles at $r_{1,2} = -0.4$. The steady state error for unit ramp reference input is

$$e_{ss} = \frac{p_c}{K_c z_c}.$$

Accordingly, the second design specification is to make the ratio $K_c z_c / p_c$ as large as possible.

The characteristic equation is

$$\chi(s) = s(s^2 + 3s + 3)(s - p_c) + K_c(s^2 - 3s + 3)(s - z_c),$$

and it is desired to be in the form

$$\chi(s) = (s + 0.4)^2(s - r_3)(s - r_4)$$

for some $r_{3,4}$ with $\mathrm{Re}(r_{3,4}) < 0$, which implies that

$$\chi(s)\Big|_{s=-0.4} = 0, \qquad \frac{d}{ds}\chi(s)\Big|_{s=-0.4} = 0. \qquad (5.13)$$

Conditions (5.13) give two equations:

$$0.784(0.4 + p_c) - 4.36K_c(0.4 + z_c) = 0$$
$$4.36K_c - 0.784 - 1.08(0.4 + p_c) + 3.8K_c(0.4 + z_c) = 0$$

from which z_c and p_c can be solved in terms of K_c. Then, by simple substitutions, the ratio to be maximized, $K_c z_c / p_c$, can be reduced to

$$\frac{K_c z_z}{p_c} = \frac{3.4776K_c - 0.784}{24.2469K_c - 3.4776}.$$

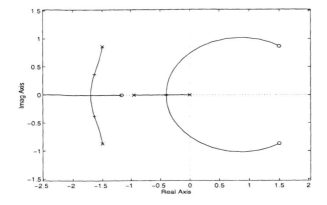

Figure 5.10: Root locus for Example 5.4.

The maximizing value of K_c is 0.1297; it leads to $p_c = -0.9508$ and $z_c = -1.1637$. For this controller, the feedback system poles are

$$\{-1.64 + j0.37, \ -1.64 - j0.37, \ -0.40, \ -0.40\}.$$

The root locus is shown in Figure 5.10.

5.2 Complementary Root Locus

In the previous section, the root locus parameter K was assumed to be positive and the phase and magnitude rules were established based on this assumption. There are some situations in which controller gain can be negative as well. Therefore, the complete picture is obtained by drawing the usual root locus (for $K > 0$) and the complementary root locus (for $K < 0$). The complementary root locus rules are

$$\ell \times 360° \ = \ \sum_{i=1}^{n} \angle(r - p_i) - \sum_{j=1}^{m} \angle(r - z_j), \quad \ell = 0, \pm1, \pm2, \ldots \ (5.14)$$

$$|K| \ = \ \frac{\prod_{i=1}^{n} |r - p_i|}{\prod_{j=1}^{m} |r - z_j|}. \tag{5.15}$$

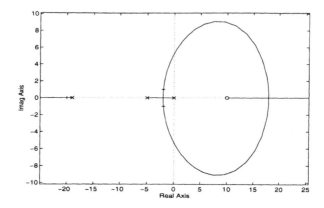

Figure 5.11: Complementary root locus for Example 5.3.

Since the phase rule (5.14) is the 180° shifted version of (5.7), the complementary root locus is obtained by simple modifications in the root locus construction rules. In particular, the number of asymptotes and their center are the same, but their angles α_ℓ's are given by

$$\alpha_\ell = \frac{2\ell}{(n-m)} \times 180°, \quad \ell = 0, \ldots, (n-m-1).$$

Also, an interval on the real axis is on the complementary root locus if and only if it is not on the usual root locus.

Example 5.5 In the Example 5.3 given above, if the problem data is modified to $p_m = -5$, $p_1 = -20$ and $p_{2,3} = -2 \pm j$, then the controller parameters become

$$K = -10 \quad p_c = -19 \quad z_c = 10.$$

Note that the gain is negative. The roots of the characteristic equation as K varies between 0 and $-\infty$ form the complementary root locus, see Figure 5.11.

Example 5.6 (Example 5.4 revisited). In this example, if K increases from $-\infty$ to $+\infty$, the closed-loop system poles move along the

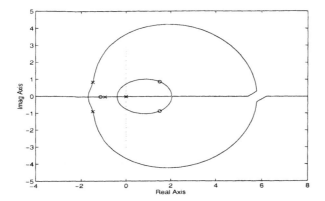

Figure 5.12: Complementary and usual root loci for Example 5.4.

complementary root locus, and then the usual root locus as illustrated in Figure 5.12.

5.3 Exercise Problems

1. Let the controller and the plant be given as

$$C(s) = \frac{K(s^2 + 0.5s + 9)}{s^2} \qquad P(s) = \frac{(s+2)}{(s^2 + 0.5s + 10)}.$$

Draw the root locus with respect to K without using any numerical tools for polynomial root solution. Show as much detail as possible.

2. Consider the feedback system with

$$C(s) = \frac{K}{(s+5)} \qquad P(s) = \frac{1}{(s-1)(s+2)}.$$

(a) Find the range of K for which the feedback system is stable.

(b) Let r_1, r_2, r_3 be the poles of the feedback system. It is desired to have

$$\mathrm{Re}(r_k) \leq -x \quad \text{for all } k = 1, 2, 3,$$

for some $x > 0$, so that the feedback system is stable. Determine the value of K that maximizes x.

Hint: Draw the root locus first.

3. (a) Draw the complementary root locus for
$$G_1(s) = \frac{(s+3)}{(s-1)(s+5)(s+4+j2)(s+4-j2)}.$$
and connect it to the root locus shown in Figure 5.3.

 (b) Let r_1, \ldots, r_4 be the roots of $1 + KG_1(s) = 0$. It is desired to have $\mathrm{Re}(r_i) \leq -\alpha$, $i = 1, \ldots 4$, for the largest possible $\alpha > 0$. Determine the value of K achieving this design goal, and show all the corresponding roots on the root locus.

4. For the plant
$$P(s) = \frac{1}{s(s+4)}$$

design a controller in the form
$$C(s) = K\frac{(s - z_c)}{(s - p_c)}$$

such that

 (i) the characteristic polynomial can be factored as
$$\chi(s) = (s^2 + 2\zeta\omega_n s + \omega_n^2)(s + r)$$
where $\zeta \in [0.6, 1]$, $\zeta\omega_n = 1$ and $r \geq 5$,

 (ii) the ratio Kz_c/p_c is as large as possible.

Draw the root locus for this system.

5. Consider the plant
$$P(s) = \frac{(s+2)}{(s^2 + 2s + 2)}.$$

Design a controller in the form

$$C(s) = \frac{K(s-z)}{(s^2 + as + b)} \;, \quad \text{where } K, z, a, b \text{ are real numbers}$$

such that

- the feedback system is stable with four closed-loop poles satisfying $r_1 = \bar{r}_2 = r_3 = \bar{r}_4$,
- the steady state tracking error is zero for $r(t) = \sin(t)\mathbb{U}(t)$.

Draw the root locus for this system and show the location of the roots for the selected controller gain.

6. Consider the plant

$$P(s) = \frac{(1 - \tau s)}{s\,(1 + \tau s)}$$

where τ is an uncertain parameter. Determine the values of τ for which the PI controller

$$C(s) = 5 + \frac{10}{s}$$

stabilizes the system. Draw the closed-loop system poles as τ varies from 0 to $+\infty$. Find the values of τ such that the dominant closed-loop poles have damping coefficient $\zeta \geq 0.85$. What is the largest possible ζ and the corresponding τ?

Useful MATLAB commands are roots, rlocus, rlocfind, and sgrid.

Chapter 6

Frequency Domain Analysis Techniques

Stability of the standard feedback system, Figure 2.11, is determined by checking whether the roots of $1 + G(s) = 0$ are in the open left half plane or not, where $G(s) = P(s)C(s)$ is the given open-loop transfer function. Suppose that $G(s)$ has finitely many poles, then it can be written as $G = N_G/D_G$, where $N_G(s)$ has no poles and $D_G(s)$ is a polynomial containing all the poles of $G(s)$. Clearly,

$$1 + G(s) = 0 \quad \Longleftrightarrow \quad F(s) := \frac{D_G(s) + N_G(s)}{D_G(s)} = 0.$$

The zeros of $F(s)$ are the closed-loop system poles, while the poles of $F(s)$ are the open-loop system poles. The feedback system is stable if and only if $F(s)$ has no zeros in $\overline{\mathbb{C}}_+$. The Nyquist stability test uses Cauchy's Theorem to determine the number of zeros of $F(s)$ in $\overline{\mathbb{C}}_+$. This is done by counting the number of encirclements of the origin by the closed path $F(j\omega)$ as ω increases from $-\infty$ to $+\infty$. Cauchy's theorem (or Nyquist stability criterion) not only determines stability

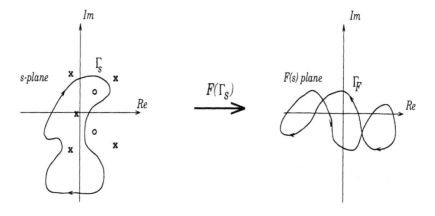

Figure 6.1: Mapping of contours.

of a feedback system, but also gives quantitative measures on *stability robustness* with respect to certain types of uncertainty.

6.1 Cauchy's Theorem

A closed path in the complex plane is a *positive* contour if it is in the *clockwise* direction. Given an analytic function $F(s)$ and a contour Γ_s in \mathbb{C}, the contour Γ_F is defined as the map of Γ_s under $F(\cdot)$, i.e. $\Gamma_F := F(\Gamma_s)$. See Figure 6.1. The contour Γ_F is drawn on a new complex plane called $F(s)$ plane. The number of encirclements of the origin by Γ_F is determined by the number of poles and zeros of $F(s)$ encircled by Γ_s. The exact relationship is given by Cauchy's theorem.

Theorem 6.1 (Cauchy's Theorem) *Assume that a positive contour Γ_s does not go through any pole or zero of $F(s)$ in the s-plane. Then, $\Gamma_F := F(\Gamma_s)$ encircles the origin of the $F(s)$ plane*

$$n_o = n_z - n_p$$

times in the positive direction, where n_z and n_p are the number of zeros and poles (respectively) of $F(s)$ encircled by Γ_s.

For a proof of this theorem see e.g., [37, pp. 584–587]. In Figure 6.1, Γ_s encircles $n_z = 1$ zero and $n_p = 2$ poles of $F(s)$; the number of positive encirclements of the origin by the contour Γ_F is $n_o = 1-2 = -1$, i.e., the number of encirclements of the origin in the counterclockwise direction is $1 = -n_o$.

6.2 Nyquist Stability Test

By using Cauchy's theorem, stability of the feedback system can be determined as follows:

(i) Define a contour Γ_s encircling the right half plane in the clockwise direction.

(ii) Obtain $\Gamma_F = 1 + \Gamma_G$ and count its positive (clockwise) encircle-ments of the origin. Let this number be denoted by n_o.

(iii) If the number of poles of $G(s)$ in Γ_s is n_p, then the number of zeros of $F(s)$ in Γ_s (i.e., the number of right half plane poles of the feedback system) is $n_z = n_o + n_p$.

In conclusion, *the feedback system is stable if and only if* $n_z = 0 \iff n_o = -n_p$, *which means that the map Γ_G encircles the point -1, in $G(s)$ plane, n_p times in the counterclockwise direction.* This stability test is known as the Nyquist stability criterion.

If $G(s)$ has no poles on the Im-axis, Γ_s is defined as:

$$\Gamma_s := \lim_{R \to \infty} \mathcal{I}_R \cup \mathcal{C}_R$$

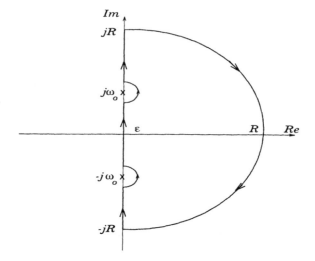

Figure 6.2: Definition of Γ_s.

where

$$\mathcal{I}_R := \{j\omega \; : \; \omega \text{ increases from } -R \text{ to } +R\}$$

$$\mathcal{C}_R := \{Re^{j\theta} \; : \; \theta \text{ decreases from } \frac{+\pi}{2} \text{ to } \frac{-\pi}{2}\}.$$

When $G(s)$ has an imaginary axis pole, say at $j\omega_o$, a small neighborhood of $j\omega_o$ is excluded from \mathcal{I}_R and a small semicircle, $\mathcal{C}_\varepsilon(j\omega_o)$, is added to connect the paths on the imaginary axis:

$$\mathcal{C}_\varepsilon(j\omega_o) := \{j\omega_o + \varepsilon e^{j\theta} \; : \; \theta \text{ increases from } \frac{-\pi}{2} \text{ to } \frac{+\pi}{2}\},$$

where $\varepsilon \to 0$. See Figure 6.2.

Since Γ_s is symmetric around the real axis and $\overline{G(s)} = G(\bar{s})$, the closed path Γ_G is symmetric around the real axis. Therefore, once $\Gamma_G^+ := G(\Gamma_s^+)$ is drawn for

$$\Gamma_s^+ := \Gamma_s \cap \{s \in \mathbb{C} \; : \; \text{Im}(s) \geq 0\},$$

the complete path Γ_G is obtained by

$$\Gamma_G = \Gamma_G^+ \cup \overline{\Gamma_G^+}\,.$$

Moreover, in most practical cases, $G(s) = P(s)C(s)$ is strictly proper, meaning that

$$\lim_{R \to \infty} G(Re^{j\theta}) = 0 \quad \forall\, \theta.$$

In such cases, Γ_G^+ is simply the path of $G(j\omega)$ as ω increases from 0 to $+\infty$, excluding Im-axis poles. This path of $G(j\omega)$ is called the Nyquist plot; in MATLAB it is generated by the nyquist command.

Example 6.1 The open-loop system transfer function considered in Section 5.1.1 is in the form

$$G(s) = K\frac{(s+3)}{(s-1)(s+5)(s+4+j2)(s+4-j2)}, \tag{6.1}$$

and we have seen that the feedback system is stable for $K \in (\frac{100}{3}, 215.83)$. In particular, for $K = 100$ the feedback system is stable. This result can be tested by counting the number of encirclements of the critical point, -1, by the Nyquist plot of $G(j\omega)$. To get an idea of the general shape of the Nyquist plot, first put $s = j\omega$ in (6.1); this gives $G(j\omega) = \frac{N_G(j\omega)}{D_G(j\omega)}$. Then multiply $N_G(j\omega)$ and $D_G(j\omega)$ by the complex conjugate of $D_G(j\omega)$ so that the common denominator for the real and imaginary parts of $G(j\omega)$ is real and positive:

$$G(j\omega) = \frac{-100(\ (9\omega^4 + 101\omega^2 + 300)\ -\ j\omega(\omega^4 - 11\omega^2 - 220)\)}{(\omega^4 - 4\omega^2 - 100)^2 + \omega^2(12\omega^2 - 40)^2}$$

Clearly, the real part is negative for all $\omega > 0$ and the imaginary part is zero at $\omega_1 = 0$ and at $\omega_2 = 4.6173$. Note that

$$\mathrm{Im}(G(j\omega)) \begin{cases} \leq 0 & \text{for } 0 \leq \omega \leq 4.6173 \\ \geq 0 & \text{for } \omega \geq 4.6173. \end{cases}$$

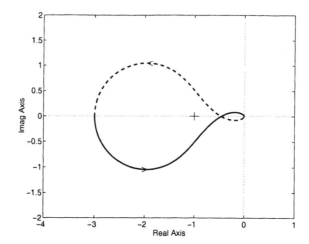

Figure 6.3: Nyquist plot of $\frac{100(s+3)}{(s-1)(s+5)(s+4+j2)(s+4-j2)}$.

Also, $G(j0) = -3$ and $G(j4.6173) = -0.4633$. These numerical values
can be used for roughly sketching the Nyquist plot. The exact plot
is shown in Figure 6.3, where the dashed line is $G(j\omega)$ for $\omega < 0$ and
the arrows show the increasing direction of ω from $-\infty$ to $+\infty$. Since
Γ_s contains one pole in \mathbb{C}_+ and Γ_G encircles -1 once in the counter
clockwise direction, the feedback system is stable. By looking at the
real axis crossings of Γ_G, it is deduced that the feedback system remains
stable for all open-loop transfer functions $\alpha G(s)$ as long as $\frac{1}{3} < \alpha <$
$\frac{1}{0.4633} \approx 2.1583$. This is consistent with the results of Section 5.1.1.

Example 6.2 The Nyquist plot for the Example 5.3 (part b) can be
obtained as follows. The open-loop transfer function is

$$G(s) = \frac{8.25\,(s + 0.73)}{s\,(s + 0.5)\,(s + 5.5)}.$$

The system is strictly proper, so $G(Re^{j\theta}) = 0$ for all θ as $R \to \infty$. In
this case, there is a pole on the imaginary axis, at $s = 0$, so Γ_s should
exclude it. The Nyquist plot will be obtained by drawing $G(s)$ for

(i) $s = \varepsilon e^{j\theta}$ with $\varepsilon \to 0$ and θ varying from 0 to $\frac{\pi}{2}$, and for (ii) $s = j\omega$ where ω varying from ε to $+\infty$. Simple substitutions give

$$G(\varepsilon e^{j\theta}) \approx \frac{2.19}{\varepsilon} e^{-j\theta}$$

and as θ varies from 0 to $\frac{\pi}{2}$ this segment of the Nyquist plot follows the quarter circle which starts at $\frac{2.19}{\varepsilon}$ and ends at $-j\frac{2.19}{\varepsilon}$. For the second segment of the Nyquist plot let $s = j\omega$ and as before, separate real and imaginary parts of $G(j\omega)$:

$$G(j\omega) = -8.25 \, \frac{\omega \, (1.63 + \omega^2) + j(5.27\omega^2 + 2.0075)}{\omega \, ((2.75 - \omega^2)^2 + 36\omega^2)}$$

Note that as $\omega \to \varepsilon$, where $0 < \varepsilon \ll 1$, we have

$$G(j\varepsilon) \approx -1.7782 - j\frac{2.19}{\varepsilon}.$$

Also, the real and imaginary parts of $G(j\omega)$ are negative for all $\omega > 0$; so this segment of the Nyquist plot follows a path which remains in the third quadrant of the complex plane. Hence, the critical point, -1, is not encircled by Γ_G. This implies that the feedback system is stable, because $G(s)$ has no poles encircled in Γ_s. The second segment of the Nyquist plot is shown in Figure 6.4.

6.3 Stability Margins

Suppose that for a given plant P and a controller C the feedback system is stable. Stability robustness, with respect to certain special types of perturbations in $G = PC$, can be determined from the Nyquist plot. Three different types of perturbations will be studied here

- gain scaling
- phase shift
- combined gain scaling and phase shift.

Throughout this section it is assumed that $G(s)$ is strictly proper.

Figure 6.4: Nyquist plot of $\frac{8.25(s+0.73)}{s(s+0.5)(s+5.5)}$.

Case 1: *G has no poles in the open right half plane.*

In this case, the closed-loop system is stable if and only if $G(j\omega)$ does not encircle the critical point, -1. A generic Nyquist plot corresponding to this class of systems is shown in Figure 6.5, (only the segment $\omega \geq 0$ is shown here).

The number of encirclements of the critical point does not change for all Nyquist plots in the form $kG(j\omega)$ as long as

$$0 < k < \frac{1}{\alpha},$$

where $-\alpha \in (-1 , 0)$ is the Re-axis crossing point closest to -1.

Similarly, the critical point is not encircled for all $e^{-j\phi}G(j\omega)$ provided

$$0 < \phi < \varphi = \min\{\pi + \angle G(j\omega_c) : \omega_c \text{ is such that } |G(j\omega_c)| = 1 \}.$$

Here we assumed that $\angle G(j\omega_c) \in (-\pi , \pi)$. Also, if $|G(j\omega)| < 1$ for all ω, then we define $\varphi := \infty$. In the light of these observations, the following definitions are made:

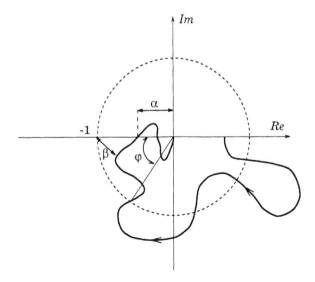

Figure 6.5: Stability margins for $G(s)$ with no pole in \mathbb{C}_+.

GM: the gain margin is $\frac{1}{\alpha} \equiv -20\log_{10}(\alpha)$ dB.

PM: the phase margin is φ (in radians).

There might be some cases in which negative real axis crossings occur, not only between 0 and -1 but also between -1 and $-\infty$, yet the feedback system is stable. For example, consider the Nyquist plot shown in Figure 6.6. The total number of encirclements of the critical point is zero (once in counterclockwise and once in the clockwise direction) and this does not change for all $kG(j\omega)$ when

$$k \in (0 \, , \, \frac{1}{\alpha_3}) \cup (\frac{1}{\alpha_2} \, , \, \frac{1}{\alpha_1}).$$

The nominal value of the gain $k = 1$ falls in the second interval, so the *upper and lower gain margins* are defined as

$$\mathrm{GM_{upp}} = \frac{1}{\alpha_1} > 1 \quad \text{and} \quad \mathrm{GM_{low}} = \frac{1}{\alpha_2} < 1.$$

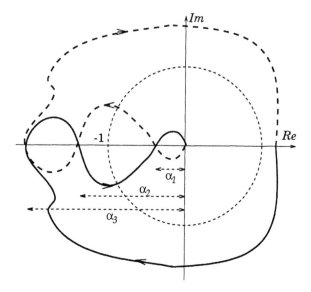

Figure 6.6: Upper and lower gain margins.

In such cases we can define the *relative gain margin* as

$$\mathrm{GM_{rel}} := \frac{\mathrm{GM_{upp}}}{\mathrm{GM_{low}}} = \frac{\alpha_2}{\alpha_1} \equiv 20 \log_{10}\left(\frac{\alpha_2}{\alpha_1}\right) \text{ dB}.$$

For good stability robustness to perturbations in the gain, we want $\mathrm{GM_{upp}}$ to be large and $\mathrm{GM_{low}}$ to be small.

Exercise: Draw the root locus for

$$G(s) = \frac{K(s+0.25)^3}{(s+0.001)^2(s+0.1)(s^2+0.5s+0.125)}, \qquad (6.2)$$

and show that for $K = 0.0536$ the system is stable. Draw the Nyquist plot for this value of K and determine the upper, lower, and relative gain margins and the phase margin.

Answer: $\mathrm{GM_{upp}} \approx 6$ dB and PM $\approx 4°$.

Figure 6.5 suggests that the distance between the critical point and the Nyquist path can be dangerously small, yet the gain and phase

margins may be large. In this situation, simultaneous perturbations in gain and phase of $G(j\omega)$ may change the number of encirclements, and hence may destabilize the system. Therefore, the most meaningful stability margin is the so-called *vector margin*, which is defined as the smallest distance between $G(j\omega)$ and -1, i.e.

$$\mathbf{VM} = \beta = \inf_{\omega} |G(j\omega) - (-1)|. \tag{6.3}$$

An upper bound for the VM can be obtained from the PM:

$$\beta = \inf_{\omega} |1 + G(j\omega)| \leq |1 + G(j\omega_c)| = 2\sin(\frac{\varphi}{2}).$$

Simple trigonometric identities give the last equality from the definition of φ in Figure 6.5. Note that, if φ is small, then β is small. A rather obvious upper bound for the VM can be determined from Figure 6.6:

$$\beta \leq \min\{ |1 - \alpha_1| , |1 - \alpha_2| \}.$$

Recall that the sensitivity function is defined as $S(s) = (1+G(s))^{-1}$. So, by using the notation of Chapter 4 (system norms) we have

$$\beta^{-1} = \sup_{\omega} |S(j\omega)| = \|S\|_{\mathcal{H}^{\infty}} . \tag{6.4}$$

In other words, vector margin is the inverse of the \mathcal{H}^{∞} norm of the sensitivity function. Hence, the VM can be determined via (6.4) by plotting $|S(j\omega)|$ and finding its peak value β^{-1}. The feedback system has "good" stability robustness (in the presence of mixed gain and phase perturbations in $G(j\omega)$) if the vector margin is "large", i.e., the \mathcal{H}^{∞} norm of the sensitivity function S is "small." Since $G(s)$ is assumed to be strictly proper $G(\infty) = 0$ and hence

$$\lim_{\omega \to \infty} |S(j\omega)| = 1.$$

Thus, $\|S\|_{\infty} \geq 1$ and $\beta \leq 1$ whenever $G(s)$ is strictly proper.

Case 2. *G has poles in the open right half plane.*

Again, suppose that the feedback system is stable. The only difference from the previous case is the generic shape of the Nyquist plot; it encircles -1 in the counterclockwise direction as many times as the number of open right half plane poles of G. Otherwise, the basic definitions of gain, phase, and vector margins are the same:

- the gain margin is the smallest value of $k > 1$ for which the feedback system becomes unstable when $G(j\omega)$ is replaced by $kG(j\omega)$.

- the phase margin is the smallest phase lag $\phi > 0$ for which the feedback system becomes unstable when $G(j\omega)$ is replaced by $e^{-j\phi}G(j\omega)$.

- the vector margin β is the smallest distance between $G(j\omega)$ and the critical point -1.

The upper, lower and relative gain margins are defined similarly.

Exercise: Consider the system (6.1) with $K = 100$. Show that the phase margin is approximately $24°$, the gain margin (i.e., the upper gain margin) is $2.1583 \equiv 6.7$ dB and the relative gain margin is $6.47 \equiv 16.2$ dB. The vector margin for this system is 0.339.

6.4 Stability Margins from Bode Plots

The Nyquist plot shows the path of $G(j\omega)$ on the complex plane as ω increases from 0 to $+\infty$. This can be translated into Bode plots, where the magnitude and the phase of $G(j\omega)$ are drawn as functions of ω. Usually, logarithmic scale is chosen for the frequency axis, and the magnitude is expressed in decibels and the phase is expressed in degrees. In MATLAB, Bode plots are generated by the bode command. The gain and phase margins can be read from the Bode plots, but the

vector margin is not apparent. We need to draw the Bode magnitude plot of $S(j\omega)$ to determine the vector margin.

Suppose that $G(s)$ has no poles in the open right half plane. Bode plots of a typical $G(j\omega)$ are shown in Figure 6.7. The gain and phase margins are illustrated in this figure:

$$\begin{aligned} \text{GM} &= -20\log_{10}|G(j\omega_p)| \\ \text{PM} &= \pi + \angle G(j\omega_c) \end{aligned}$$

where ω_c is the gain crossover frequency, $20\log_{10}|G(j\omega_c)| = 0$ dB, and ω_p is the phase crossover frequency, $\angle G(j\omega_p) = -180°$. The upward arrows in the figure indicate that the gain and phase margins of this system are positive, hence the feedback system is stable. Note how the arrows are drawn: on the gain plot the arrow is drawn from the magnitude of G at ω_p to 0 dB and on the phase plot from $-180°$ to $\angle G(j\omega_c)$. If either GM, or PM, or both, are negative then the closed-loop system is unstable. The gain and phase margins can be obtained via MATLAB by using the `margin` command.

Example 6.3 Consider the open-loop system

$$G(s) = \frac{0.1297\ (s^2 - 3s + 3)\ (s + 1.1637)}{s\ (s^2 + 3s + 3)\ (s + 0.9508)},$$

which is designed in example 4 of Section 5.1.2. The Bode plots given in Figure 6.8 show that the phase margin is about 70° and the gain margin is approximately 14 dB.

Exercise: Draw the Bode plots for $G(s)$ given in (6.2) with $K = 0.0536$. Determine the gain and phase margins from the `margin` command of MATLAB. Verify the results by comparing them with the stability margins obtained from the Nyquist plot.

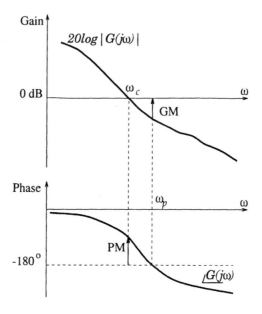

Figure 6.7: Gain and phase margins from Bode plots.

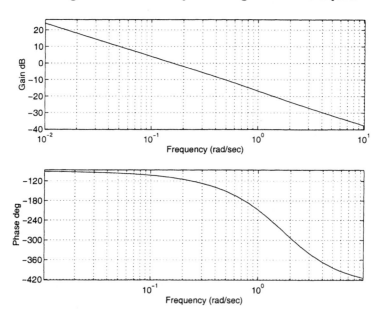

Figure 6.8: Bode plots of $\dfrac{0.1297\ (s^2-3s+3)\ (s+1.1637)}{s\ (s^2+3s+3)\ (s+0.9508)}$.

6.5 Exercise Problems

1. Sketch the Nyquist plots of

 (a) $G(s) = \dfrac{K(s+1)}{s(s-1)}$

 (b) $G(s) = \dfrac{K(s+3)}{(s^2+16)}$

 and compute the gain and phase margins as functions of K.

2. Sketch the Nyquist plot of

 $$G(s) = \frac{K(s+5)(s^2+0.3s+2)}{s(s+4)(s^2+4)}$$

 and show that the feedback system is stable for all $K > 0$. By using root locus, find K, which places the closed-loop system poles to the left of $\mathrm{Re}(s) = -\sigma$ for the largest possible $\sigma > 0$. What is the vector margin of this system?

3. Consider the system defined by

 $$G(s) = \frac{K(s^2 - 3s + 3)(s - z_c)}{s(s^2 + 3s + 3)(s - p_c)}$$

 where $K = 1$, $p_c = 3z_c$ and z_c is the parameter to be adjusted.

 (a) Find the range of z_c for which the feedback system is stable.

 (b) Design z_c so that the vector margin is as large as possible.

 (c) Sketch the Nyquist plot and determine the gain and phase margins for the value of z_c computed in part (b).

4. For the open-loop system

 $$G(s) = \frac{K(s+3)}{(s-1)(s+5)(s^2+8s+20)}$$

 it was shown that the feedback system is stable for all $K \in (33.33 , 215.83)$.

(a) Find the value of K that maximizes the quantity

$$\min\{\text{GM}_{\text{upp}}, \ (\text{GM}_{\text{low}})^{-1}\}.$$

(b) Find the value of K that maximizes the *phase margin*.

(c) Find the value of K that maximizes the *vector margin*.

Chapter 7

Systems with Time Delays

Mathematical models of systems with time delays were introduced in Section 2.2.2. If there is a time delay in a linear system, then its transfer function includes a term e^{-hs} where $h > 0$ is the delay time. For this class of systems, transfer functions cannot be written as ratios of two polynomials of s. The discussion on Routh-Hurwitz and Kharitanov stability tests, and root locus techniques do not directly extend to systems with time delays. In order to be able to apply these methods, the delay element e^{-hs} must be approximated by a rational function of s. For this purpose, Padé approximations of delay systems will be examined in this chapter.

The Nyquist stability criterion remains valid for a large class of delay systems, so there is no need for approximations in this case. Several examples will be given to illustrate the effects of time delay on stability margins, and the concept of *delay margin* will be introduced shortly.

The standard feedback control system considered in this chapter is shown in Figure 7.1, where the controller C and plant P are in the form

$$C(s) = \frac{N_c(s)}{D_c(s)}$$

Figure 7.1: Feedback system with time delay.

and

$$P(s) = e^{-hs}P_0(s) \quad \text{where} \quad P_0(s) = \frac{N_p(s)}{D_p(s)}$$

with (N_c, D_c) and (N_p, D_p) being coprime pairs of polynomials. The open-loop transfer function is

$$G(s) = e^{-hs}G_0(s),$$

where $G_0(s) = P_0(s)C(s)$ corresponds to the case $h = 0$.

The response of a delay element e^{-hs} to an input $u(t)$ is simply $u(t - h)$, which is the h units of time delayed version of $u(t)$. Hence by definition, the delay element is a stable system. In fact, the $\mathcal{L}_p[0,\infty)$ norm of its output is equal to the $\mathcal{L}_p[0,\infty)$ norm of its input for any p. Thus, the system norm (defined as the induced $\mathcal{L}_p[0,\infty)$ norm) of e^{-hs} is unity.

The function e^{-hs} is analytic in the entire complex plane and has no finite poles or zeros. The frequency response of the delay element is determined by its magnitude and phase on the Im-axis:

$$|e^{-jh\omega}| = 1 \quad \text{for all} \quad \omega$$
$$\angle e^{-jh\omega} = -h\omega .$$

The magnitude identity confirms the norm preserving property of the delay element. Moreover, it also implies that the transfer function e^{-hs} is *all-pass*. For $\omega > 0$ the phase is negative and it is linearly decreasing.

The Bode and Nyquist plots of $G(j\omega)$ are determined from the identities

$$|G(j\omega)| \;=\; |G_0(j\omega)| \tag{7.1}$$

$$\angle G(j\omega) \;=\; -h\omega + \angle G_0(j\omega) \,. \tag{7.2}$$

This fact will be used later, when stability margins are discussed.

7.1 Stability of Delay Systems

Stability of the feedback system shown in Figure 7.1 is equivalent to having all the roots of

$$\chi(s) := D_c(s)D_p(s) + e^{-hs} N_c(s)N_p(s) \tag{7.3}$$

in the open left half plane, \mathbb{C}_-. Strictly speaking, $\chi(s)$ is not a polynomial because it is a transcendental function of s. The functions of the form (7.3) belong to a special class of functions called *quasi-polynomials*. Recall that the plant and the controller are causal systems, so their transfer functions are proper: $deg(N_c) \leq deg(D_c)$ and $deg(N_p) \leq deg(D_p)$. In fact, most physical plants are strictly proper, accordingly assume that $deg(N_p) < deg(D_p)$.

The closed-loop system poles are the roots of

$$1 + G(s) = 0 \quad \Longleftrightarrow \quad 1 + e^{-hs}G_0(s) = 0 \,,$$

or the roots of

$$\chi(s) = D(s) + e^{-hs} N(s) \tag{7.4}$$

where $D(s) = D_c(s)D_p(s)$ and $N(s) = N_c(s)N_p(s)$. Hence, to determine closed-loop system stability, we need to check that the roots of (7.4) are in \mathbb{C}_-. Following are known facts (see [8, 49])

(i) if r_k is a root of (7.4), then so is \bar{r}_k, (i.e., roots appear in complex conjugate pairs as usual),

(ii) there are infinitely many poles $r_k \in \mathbb{C}$, $k = 1, 2, \ldots$, satisfying $\chi(r_k) = 0$,

(iii) and r_k's can be enumerated in such a way that $\mathrm{Re}(r_{k+1}) \le \mathrm{Re}(r_k)$; moreover, $\mathrm{Re}(r_k) \to -\infty$ as $k \to \infty$.

Example 7.1 If $G(s) = e^{-hs}/s$, then the closed-loop system poles r_k, for $k = 1, 2, \ldots$, are the roots of

$$1 + \frac{e^{-h\sigma_k} e^{-jh\omega_k}}{\sigma_k + j\omega_k} e^{\pm j2k\pi} = 0 \tag{7.5}$$

where $r_k = \sigma_k + j\omega_k$ for some $\sigma_k, \omega_k \in \mathbb{R}$. Note that $e^{\pm j2k\pi} = 1$ for all $k = 1, 2, \ldots$. The equation (7.5) is equivalent to the following set of equations

$$e^{-h\sigma_k} = |\sigma_k + j\omega_k| \tag{7.6}$$

$$\pm(2k - 1)\pi = h\omega_k + \angle(\sigma_k + j\omega_k) \quad k = 1, 2, \ldots \tag{7.7}$$

It is quite interesting that for $h = 0$ there is only one root $r = -1$, but even for infinitesimally small $h > 0$ there are infinitely many roots. From the magnitude condition (7.6), it can be shown that

$$\sigma_k \ge 0 \quad \Longrightarrow \quad |\omega_k| \le 1. \tag{7.8}$$

Also, for $\sigma_k \ge 0$, the phase $\angle(\sigma_k + j\omega_k)$ is between $\frac{-\pi}{2}$ and $\frac{+\pi}{2}$, therefore (7.7) leads to

$$\sigma_k \ge 0 \quad \Longrightarrow \quad h\,|\omega_k| \ge \frac{\pi}{2}. \tag{7.9}$$

By combining (7.8) and (7.9), it can be proven that the feedback system has no roots in the closed right half plane when $h < \frac{\pi}{2}$. Furthermore, the system is unstable if $h \ge \frac{\pi}{2}$. In particular, for $h = \frac{\pi}{2}$ there are two

roots on the imaginary axis, at $\pm j1$. It is also easy to show that, for any $h > 0$ as $k \to \infty$, the roots converge to

$$r_k \to \frac{1}{h}\left(-\ln(\frac{2k\pi}{h}) \pm j2k\pi\right).$$

As $h \to 0$, the magnitude of the roots converge to ∞.

As illustrated by the above example, property (iii) implies that for any given real number σ there are only finitely many r_k's in the region of the complex plane

$$\mathbb{C}_\sigma := \{s \in \mathbb{C} \; : \; \text{Re}(s) \geq \sigma\}.$$

In particular, with $\sigma = 0$, this means that the quasi-polynomial $\chi(s)$ can have only finitely many roots in the right half plane. Since the effect of the closed-loop system poles that have very large negative real parts is negligible (as far as closed-loop systems' input–output behavior is concerned), only finitely many "dominant" roots r_k, for $k = 1, \ldots, m$, should be computed for all practical purposes.

7.2 Padé Approximation of Delays

Consider the following model-matching problem for a strictly proper *stable* rational transfer function $G_0(s)$: approximate $G(s) = e^{-hs}G_0(s)$ by $\widehat{G}(s) = P_d(s)G_0(s)$, where $P_d(s) = N_d(s)/D_d(s)$ is a rational approximation of the time delay. We want to choose $P_d(s)$ so that the input–output behavior of $\widehat{G}(s)$ matches the input-output behavior of $G(s)$. To measure the mismatch, apply the same input $u(t)$ to both $G(s)$ and $\widehat{G}(s)$. Then, by comparing the respective outputs $y(t)$ and $\widehat{y}(t)$, we can determine how well \widehat{G} approximates G, see Figure 7.2.

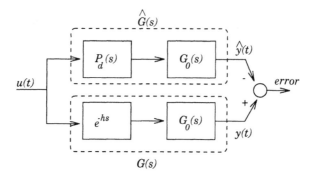

Figure 7.2: Model-matching problem.

The model-matching error (MME) will be measured by

$$\text{MME} := \sup_{u \neq 0} \frac{\|y - \widehat{y}\|_2}{\|u\|_2} \tag{7.10}$$

where $\|y - \widehat{y}\|_2$ denotes the energy of the output error due to an input with energy $\|u\|_2$. The largest possible ratio of the output error energy over the input energy is defined to be the model-matching error. From the system norms defined in Chapter 4, $\text{MME} = \text{MME}_{\mathcal{H}\infty} = \text{MME}_{\mathcal{L}\infty}$:

$$\text{MME}_{\mathcal{H}\infty} \;=\; \|G - \widehat{G}\|_{\mathcal{H}\infty} \tag{7.11}$$

$$\text{MME}_{\mathcal{L}\infty} \;=\; \sup_{\omega} |G(j\omega) - \widehat{G}(j\omega)|$$

$$\;=\; \sup_{\omega} |G_0(j\omega)| \, |(e^{-jh\omega} - P_d(j\omega))|. \tag{7.12}$$

It is clear that if $\text{MME}_{\mathcal{L}\infty}$ is small, then the difference between the Nyquist plots of $G(j\omega)$ and $\widehat{G}(j\omega)$ is small. This observation is valid even if $G_0(s)$ is *unstable*. Thus, for a given $G_0(s)$ (which may or may not be stable), we want to find a rational approximation $P_d(s)$ for the delay term e^{-hs} so that the approximation error $\text{MME}_{\mathcal{L}\infty}$ is smaller than a specified tolerance, say $\delta > 0$.

Padé approximation will be used here:

$$e^{-hs} \approx P_d(s) = \frac{N_d(s)}{D_d(s)} = \frac{\sum_{k=0}^{n} (-1)^k c_k h^k s^k}{\sum_{k=0}^{n} c_k h^k s^k}.$$

The coefficients are

$$c_k = \frac{(2n-k)!\, n!}{2n!\, k!\, (n-k)!}, \quad k = 0, 1, \ldots, n.$$

For $n = 1$ the coefficients are $c_0 = 1$, $c_1 = 1/2$, and

$$P_d(s) = \left(\frac{1 - hs/2}{1 + hs/2}\right).$$

For $n = 2$ the coefficients are $c_0 = 1$, $c_1 = 1/2$, $c_2 = 1/12$, and the second-order approximation is

$$P_d(s) = \left(\frac{1 - hs/2 + (hs)^2/12}{1 + hs/2 + (hs)^2/12}\right).$$

Now we face the following model order selection problem: Given h and $G_0(s)$, what should be the degree of the Padé approximation, n, so that the the error $\text{MME}_{\mathcal{L}\infty}$ is less than or equal to a given error bound $\delta > 0$?

Theorem 7.1 [33] *The approximation error satisfies*

$$|(e^{-jh\omega} - P_d(j\omega))| \leq \begin{cases} 2(\frac{eh\omega}{4n})^{2n+1} & \omega \leq \frac{4n}{eh} \\ 2 & \omega \geq \frac{4n}{eh} \end{cases}$$

In the light of Theorem 7.1, we can solve the model order selection problem using the following procedure:

1. From the magnitude plot of $G_0(j\omega)$ determine the frequency ω_x such that

$$|G_0(j\omega)| \leq \frac{\delta}{2} \quad \text{for all} \quad \omega \geq \omega_x$$

 and initialize $n = 1$.

2. For each $n \geq 1$ define

$$\omega_n = \max\{\omega_x, \frac{4n}{eh}\}$$

and plot the function

$$
\mathcal{E}(\omega) := \begin{cases} 2|G_0(j\omega)| \left(\frac{eh\omega}{4n}\right)^{2n+1} & \text{for } \omega \leq \frac{4n}{eh} \\ 2|G_0(j\omega)| & \text{for } \omega_n \geq \omega \geq \frac{4n}{eh} \end{cases}
$$

3. Define

$$
\mathsf{E}(n) = \frac{1}{\delta} \ \max\{\mathcal{E}(\omega) \ : \ \omega \in [0\,,\omega_x]\}.
$$

If $\mathsf{E}(n) \leq 1$, stop, this value of n satisfies the desired error bound: $\mathrm{MME}_{\mathcal{L}^\infty} \leq \delta$. Otherwise, increase n by 1, and go to Step 2.

4. Plot the approximation error function

$$
|G_0(j\omega)| \ |(e^{-jh\omega} - P_d(j\omega))|
$$

and verify that its peak value is less than δ.

Since $G_0(s)$ is strictly proper, the algorithm will pass Step 3 eventually for some finite $n \geq 1$. At each iteration, we have to draw the error function $\mathcal{E}(\omega)$ and check whether its peak is less than δ. Usually, for good results in stability and performance analysis, δ is chosen to be in the order of 10^{-2} to 10^{-4}. In most cases, as δ decreases, ω_x increases, and that forces n to increase. On the other hand, for very large values of n, the relative magnitude c_0/c_n of the coefficients of $P_d(s)$ become very large, in which case numerical difficulties arise in analysis and simulations. Also, as time delay h increases, n should be increased to keep the level of the approximation error δ fixed. This is a fundamental difficulty associated with time delay systems.

Example 7.2 Let $N(s) = (s+1)$, $D(s) = (s^2+2s+2)$ and $h = 0.1$. The magnitude plot of $G_0 = N/D$ shows that if $\delta = 0.1$, then $\omega_x = 20$, see Figure 7.3. By applying the algorithm proposed above, the normalized error $\mathsf{E}(n)$ is obtained, see Figure 7.4. Note that $\mathsf{E}(2) < 1$, which means that the choice $n = 2$ guarantees $\mathrm{MME}_{\mathcal{L}^\infty} \leq 0.1$.

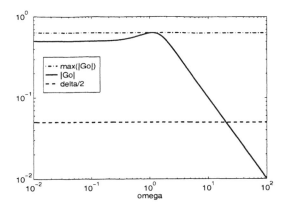

Figure 7.3: Magnitude plot of $G_0(j\omega)$.

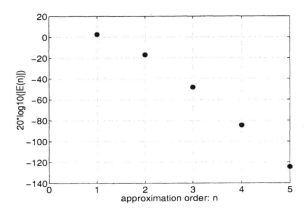

Figure 7.4: Detection of the smallest n.

Exercise. For the above example, draw the root locus of

$$D(s)D_d(s) + KN(s)N_d(s) = 0$$

for $n = 1$, 2 and 3. Show that in the region $\text{Re}(s) \geq -10$, predicted roots are approximately the same for $n = 2$ and $n = 3$, for the values of K in the interval $0 < K < 60$. Determine the range of K for which the roots remain in the left half plane for these three root loci. Comment on the results.

7.3 Roots of a Quasi-Polynomial

In this section, we discuss the following problem: given $N(s)$, $D(s)$ and $h \geq 0$, find the dominant roots of the quasi-polynomial

$$\chi(s) = D(s) + e^{-hs}N(s) .$$

For each fixed $h > 0$, it can be shown that there exists σ_{\max} such that $\chi(s)$ has no roots in the region $\mathbb{C}_{\sigma_{\max}}$, see [51] for a simple algorithm to estimate σ_{\max}, based on Nyquist criterion. Given $h > 0$ and a region of the complex plane defined by $\sigma_{\min} \leq \text{Re}(s) \leq \sigma_{\max}$, the problem is to find the roots of $\chi(s)$ in this region.

A point $r = \sigma + j\omega$ in \mathbb{C} is a root of $\chi(s)$ if and only if

$$D(\sigma + j\omega) = -e^{-h\sigma}e^{-jh\omega}N(\sigma + j\omega)$$

Taking the magnitude square of both sides of the above equation, $\chi(r) = 0$ implies

$$A_\sigma(x) := D(\sigma + x)D(\sigma - x) - e^{-2h\sigma}N(\sigma + x)N(\sigma - x) = 0$$

where $x = j\omega$. The term $D(\sigma + x)$ stands for the function $D(s)$ evaluated at $\sigma + x$. The other terms of $A_\sigma(x)$ are calculated similarly. For each fixed σ, the function $A_\sigma(x)$ is a polynomial in the variable x. By symmetry, if x is a zero of $A_\sigma(\cdot)$ then $(-x)$ is also a zero.

If $A_\sigma(x)$ has a root x_ℓ whose real part is zero, set $r_\ell = \sigma + x_\ell$. Next, evaluate the magnitude of $\chi(r_\ell)$, if it is zero, then r_ℓ is a root of $\chi(s)$. Conversely, if $A_\sigma(x)$ has no root on the imaginary axis, then $\chi(s)$ cannot have a root whose real part is the fixed value of σ from which $A_\sigma(\cdot)$ is constructed.

Algorithm: Given $N(s)$, $D(s)$, h, σ_{\min} and σ_{\max}.

Step 1. Pick σ values $\sigma_1, \ldots, \sigma_M$ between σ_{\min} and σ_{\max} such that $\sigma_{\min} = \sigma_1$, $\sigma_i < \sigma_{i+1}$ and $\sigma_M = \sigma_{\max}$. For each σ_i perform the following.

Step 2. Construct the polynomial $A_i(x)$ according to

$$A_i(x) := D(\sigma_i + x)D(\sigma_i - x) - e^{-2h\sigma_i}N(\sigma_i + x)N(\sigma_i - x)$$

Step 3. For each imaginary axis roots x_ℓ of A_i, perform the following test: Check if $|\chi(\sigma_i + x_\ell)| = 0$, if yes, then $r = \sigma_i + x_\ell$ is a root of $\chi(s)$; if not discard x_ℓ.

Step 4. If $i = M$, stop; else increase i by 1 and go to **Step 2**.

Example 7.3 Now we will find the dominant roots of

$$1 + \frac{e^{-hs}}{s} = 0. \tag{7.13}$$

We have seen that (7.13) has a pair of roots $\pm j1$ when $h = \pi/2 = 1.57$. Moreover, dominant roots of (7.13) are in the right half plane if $h > 1.57$, and they are in the left half plane if $h < 1.57$. So, it is expected that for $h \in (1.2, 2.0)$ the dominant roots are near the imaginary axis. Take $\sigma_{\min} = -0.5$ and $\sigma_{\max} = 0.5$, with $M = 400$ linearly spaced σ_i's between them. In this case

$$A_i(x) = \sigma_i^2 - e^{-2h\sigma_i} - x^2.$$

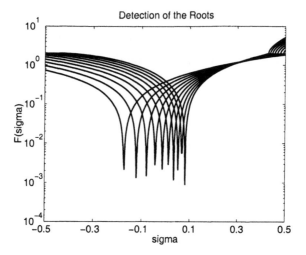

Figure 7.5: Detection of the dominant roots.

Whenever $e^{-2h\sigma_i} \geq \sigma_i^2$, $A_i(x)$ has two roots

$$x_\ell = \pm j\sqrt{e^{-2h\sigma_i} - \sigma_i^2} \qquad \ell = 1, 2.$$

For each fixed σ_i satisfying this condition, let $r_\ell = \sigma_i + x_\ell$ (note that x_ℓ is a function of σ_i, so r_ℓ is a function of σ_i) and evaluate

$$F(\sigma_i) := \left| 1 + \frac{e^{-hr_\ell}}{r_\ell} \right|.$$

If $F(\sigma_i) = 0$, then r_ℓ is a root of (7.13). For 10 different values of $h \in (1.2 , 2.0)$ the function $F(\sigma)$ is plotted in Figure 7.5. This figure shows the feasible values of σ_i for which r_ℓ (defined from σ_i) is a root of (7.13). The dominant roots of (7.13), as h varies from 1.2 to 2.0, are shown in Figure 7.6. For $h < 1.57$ all the roots are in \mathbb{C}_-. For $h > 1.57$ the dominant roots are in \mathbb{C}_+, and for $h = 1.57$ they are at $\pm j1$.

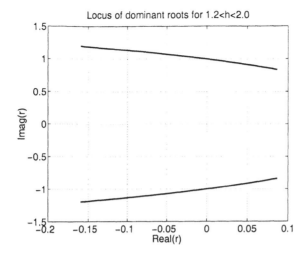

Figure 7.6: Dominant roots as h varies from 1.2 to 2.0.

7.4 Delay Margin

Consider a strictly proper system with open-loop transfer function $G(s) = e^{-hs}G_0(s)$. Suppose that the feedback system is stable when there is no time delay, i.e., the roots of

$$1 + e^{-hs}G_0(s) = 0 \tag{7.14}$$

are in \mathbb{C}_- for $h = 0$. Then, by continuity arguments, it can be shown that the feedback system with time delay $h > 0$ is stable provided h is small enough, i.e., the roots of (7.14) remain in \mathbb{C}_- for all $h \in [0 \ , \ h')$, for sufficiently small $h' > 0$. An interesting question is this: given $G_0(s)$ for which the feedback system is stable, what is h_{\max}: the largest value of h'? At the critical value of $h = h_{\max}$, all the roots of (7.14) are in the open left half plane, except finitely many roots on the imaginary axis. Therefore, the equation

$$1 + e^{-j\omega h_{\max}}G_0(j\omega) = 0 \tag{7.15}$$

holds for some real ω. Equating magnitude and phase of (7.15)

$$1 \ = \ |G_0(j\omega)|$$
$$-\pi \ = \ -\omega\, h_{max} + \angle G_0(j\omega) \ .$$

Let ω_c be the crossover frequency for this system, i.e., $|G_0(j\omega_c)| = 1$. Recall that the quantity

$$\varphi = (\pi + \angle G_0(j\omega_c))$$

is the phase margin of the system. Thus, from the above equations

$$h_{max} = \frac{\varphi}{\omega_c}.$$

If there are multiple crossover frequencies $\omega_{c1}, \ldots, \omega_{c\ell}$ for $G_0(j\omega)$ (i.e., $|G_0(j\omega_{ci})| = 1$ for all $i = 1, \ldots, \ell$), then the phase margin is

$$\varphi \ = \ \min\{\ \varphi_i \ : \ 1 \leq i \leq \ell\} \ \text{where}$$
$$\varphi_i \ = \ (\pi + \angle G_0(j\omega_{ci})).$$

In this case, h_{max} is given by

$$h_{max} = \min\{\frac{\varphi_i}{\omega_{ci}} \ : \ 1 \leq i \leq \ell\}.$$

Now consider $G(s) = e^{-hs}G_0(s)$, for some $h \in [0 \ , \ h_{max})$. In the light of the above discussion, the feedback system remains stable for all open-loop transfer functions $e^{-\tau s}G(s)$ provided

$$\tau < (h_{max} - h).$$

DM: The quantity $(h_{max} - h)$ is called the *delay margin* of the system whose open-loop transfer function is $G(s)$.

The delay margin **DM** determines how much the time delay can be increased without violating stability of the feedback system.

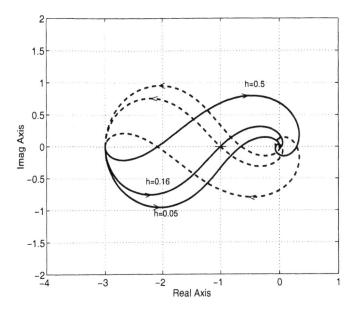

Figure 7.7: Effect of time delay (Nyquist plots).

Example 7.4 In Example 6.1, the Nyquist plot of

$$G_0(s) = \frac{100(s+3)}{(s-1)(s+5)(s+8s+20)}$$

was used to demonstrate stability of this feedback system. By using the `margin` command of MATLAB it can be shown that the phase margin for this system is $24.2° \equiv 0.42$ rad and the crossover frequency is $\omega_c = 2.66$ rad/sec. Therefore, $h_{max} = 0.42/2.66 \cong 0.16$ sec. This means that the feedback system with open-loop transfer function $G(s) = e^{-hs}G_0(s)$ is stable for all values of $h \in [0 , 0.16)$, and if $h \geq 0.16$, it is unstable. The Nyquist plots with different values of h (0.05, 0.16 and 0.5) are shown in Figure 7.7.

Example 7.5 Consider the open-loop transfer function

$$G_0(s) = \frac{10}{s(s+6)} .$$

The feedback system is stable with closed-loop poles at

$$r_{1,2} = -3 \pm j1 .$$

The Bode plots of G_0 are shown in Figure 7.8. The phase margin is $75° \equiv 1.309$ rad and the crossover frequency is $\omega_c = 1.61$ rad/sec, hence $h_{max} = 1.309/1.61 = 0.813$ sec. The Bode plots of $G(s) = e^{-hs}G_0(s)$ are also shown in the same figure; they are obtained from the relations (7.1) and (7.2). The magnitude of G is the same as the magnitude of G_0. The phase is reduced by an amount $h\omega$ at each frequency ω. The phase margin for G is

$$\varphi = 1.309 - 1.61\, h$$

For example, when $h = 0.3$ the phase margin is 0.826 rad, which is equivalent to 47°. This is confirmed by the phase plot of Figure 7.8.

Example 7.6 Consider a system with multiple crossover frequencies

$$G_0(s) = \frac{20(s^2 + 0.4s + 1)}{s^2(s^2 + 2.4s + 36)}.$$

Nyquist and Bode plots of $G_0(j\omega)$ are shown in Figures 7.9 and 7.10, respectively. The crossover frequencies are

$$\omega_{c1} = 0.57 \text{ rad/sec}, \quad \omega_{c2} = 4.64 \text{ rad/sec}, \quad \omega_{c3} = 6.57 \text{ rad/sec}.$$

with phase values

$$\angle G_0(j\omega_{c1}) = -161° , \quad \angle G_0(j\omega_{c2}) = -40° , \quad \angle G_0(j\omega_{c3}) = -127°.$$

Therefore, $\varphi_1 = 19°$, $\varphi_2 = 140°$ and $\varphi_1 = 53°$. The phase margin is

$$\varphi = \min\{\varphi_1, \varphi_2, \varphi_3\} = \varphi_1 = 19°.$$

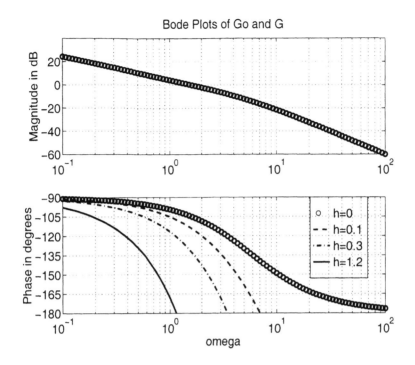

Figure 7.8: Effect of time delay (Bode plots).

However, the minimum of φ_i/ω_{ci} is achieved for $i = 3$:

$$\frac{\varphi_1}{\omega_{c1}} = 0.59 \text{ sec}, \quad \frac{\varphi_2}{\omega_{c2}} = 0.53 \text{ sec}, \quad \frac{\varphi_3}{\omega_{c3}} = 0.14 \text{ sec}.$$

Hence $h_{\max} = 0.14$ sec. Nyquist and Bode plots of $G(s) = e^{-hs}G_0(s)$ with $h = h_{\max}$ are also shown in Figures 7.9 and 7.10, respectively. From these figures, it is clear that the system is stable for all $h < h_{\max}$ and unstable for $h \geq h_{\max}$.

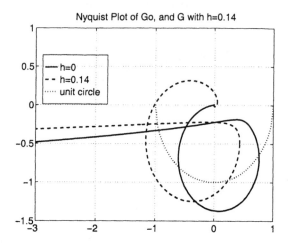

Figure 7.9: Nyquist plots for Example 7.6.

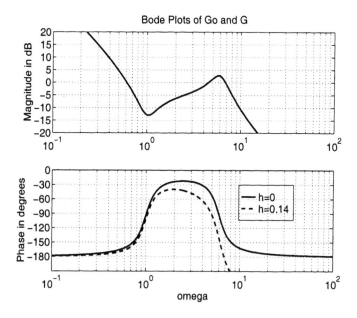

Figure 7.10: Bode plots for delay margin computation.

7.5 Exercise Problems

1. Consider the feedback system defined by

$$C(s) = \frac{0.5}{(s + 0.001)}, \quad P_0(s) = \frac{1}{s^2 + s + 1}$$

with possible time delay in the plant $P(s) = e^{-hs} P_0(s)$.

(a) Calculate h_{max}.

(b) Let $h = 1$. Using the algorithm given in Section 7.2, find the minimal order of Padé approximation such that

$$\text{MME}_{\mathcal{L}^\infty} := \|G - G_0 P_d\|_\infty \le \delta = \frac{1}{100} \ .$$

(c) Draw the locus of the dominant roots of

$$1 + G_0(s)e^{-hs} = 0$$

for $h \in (0.9 \ , \ 1.9)$.

(d) Draw the Nyquist plot of $G(s)$ for $h = 0.95, 1.55, 1.85$, and verify the result of the first problem. What is the delay margin if $h = 1$?

2. Assume that the plant is given by

$$P(s) = e^{-hs} \left(\frac{1}{s^2 + s + 1} \right),$$

where h is the amount of the time delay in the process.

(a) Design a first-order controller of the form

$$C(s) = K_c \frac{(s + z_0)}{(s + 5)}$$

such that when $h = 0$ the closed-loop system poles are equal, $r_1 = r_2 = r_3$. What is h_{max}?

(b) Let z_0 be as found in part (a), and $h = 1.2$. Estimate the location of the dominant closed-loop system poles for K_c as determined in part (a). How much can we increase K_c without violating feedback system stability?

3. Compute the delay margin of the system

$$G(s) = \frac{Ke^{-2s}(s^2 + 0.03s + 0.01)(s + 1)}{s^2(s^2 + 0.06s + 0.04)}$$

when (a) $K = 0.02$, (b) $K = 0.025$.

4. Consider the feedback system with

$$C(s) = K \quad \text{and} \quad P(s) = \frac{e^{-hs}(s + 1)}{s(s - 1)}.$$

For a fixed $K > 1$, let $h_{\max}(K)$ denote the largest allowable time delay. What is the optimal K maximizing $h_{\max}(K)$? To solve this problem by hand, you may use the approximation

$$\tan^{-1}(x) = 1.06x - 0.271x^2 \quad \text{for } 0 < x < \frac{\pi}{2}.$$

Chapter 8

Lead, Lag, and PID Controllers

In this chapter first-order lead and lag controller design principles will be outlined. Then, PID controllers will be designed from the same principles. For simplicity, the discussion here is restricted to plants with no open right half plane poles.

Consider the standard feedback system with a plant $P(s)$ and a controller $C(s)$. Assume that the plant and the controller have no poles in the open right half plane and that there are no imaginary axis pole zero cancelations in the product $G(s) = P(s)C(s)$. In this case, feedback system stability can be determined from the Bode plots of $G(j\omega)$. The design goals are:

(i) feedback system is stable with a specified phase margin, and

(ii) steady state error for unit step (or unit ramp) input is less than or equal to a specified quantity.

The second design objective determines a lower bound for the DC

gain of the open-loop transfer function, $G(0)$. In general, the larger the gain of $G(s)$, the smaller the phase margin. Therefore, we choose the DC gain as the lowest possible value satisfying (ii). Then, additional terms, with unity DC gain, are appended to the controller to take care of the first design goal. More precisely, the controller is first assumed to be a static gain $C_1(s) = K_c$, which is determined from (ii). Then, from the Bode plots of $G_1(s) = C_1(s)P(s)$ the phase margin is computed. If this simple controller does not achieve (i), the controller is modified as

$$C(s) = C_1(s)C_2(s), \quad \text{where} \quad C_2(0) = 1.$$

The DC gain of $C_2(s)$ is unity, so that the steady state error does not change with this additional term. The poles and zeros of $C_2(s)$ are chosen in such a way that the specified phase margin is achieved.

Example 8.1 For the plant

$$P(s) = \frac{1}{s(s + 0.1)}$$

design a controller such that the feedback system is stable with a phase margin of $45°$, and the steady state error for a unit ramp reference input is less than or equal to $\frac{1}{20}$. Assuming that the feedback system is stable, the steady state error for unit ramp reference input is

$$e_{ss} = \left(\lim_{s \to 0} sG(s) \right)^{-1}.$$

Hence, the steady state error requirement is satisfied if $C_1(s) = K_c \geq 2$. For $G_1(s) = K_c P(s)$ the largest phase margin is achieved with the smallest feasible K_c. Thus $K_c = 2$ and for this value of K_c the phase margin, determined from the Bode plots of G_1, is approximately $4°$. So, additional term $C_2(s)$ is needed to bring the phase margin to $45°$.

Several different choices of $C_2(s)$ will be studied in this chapter. Before discussing specific controller design methods, we make the following definitions and observations.

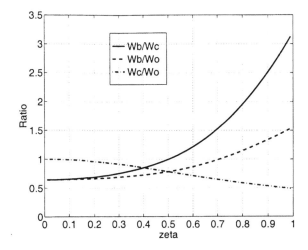

Figure 8.1: ω_b and ω_c as functions of ζ.

Consider a feedback system with closed-loop transfer function (from reference input r to system output y) $T(s) = \frac{G(s)}{1+G(s)}$ satisfying

$$|T(j\omega)| \quad \begin{cases} \geq \frac{1}{\sqrt{2}} \equiv -3 \text{ dB} & \text{for all } \omega \leq \omega_b \\ < \frac{1}{\sqrt{2}} \equiv -3 \text{ dB} & \text{for all } \omega > \omega_b \end{cases}$$

The *bandwidth* of the system is ω_b rad/sec. Usually, the bandwidth, ω_b, and the gain crossover frequency ω_c (where $|G(j\omega_c)| = 1$) are of the same order of magnitude. For example, let

$$G(s) = \frac{\omega_o^2}{s(s + 2\zeta\omega_o)} \quad \text{then} \quad T(s) = \frac{\omega_o^2}{s^2 + 2\zeta\omega_o s + \omega_o^2} \quad .$$

By simple algebraic manipulations it can be shown that both ω_b and ω_c are in the form

$$\omega_{b,c} = \omega_o \sqrt{\sqrt{1 + x_{b,c}^2} - x_{b,c}}$$

where $x_c = 2\zeta^2$ for ω_c, and $x_b = (1 - 2\zeta^2)$ for ω_b. See Figure 8.1.

Note that as $\zeta\omega_o$ increases, ω_b increases and vice versa. On the other hand, $\zeta\omega_o$ is inversely proportional to the settling time of the step response. Therefore, we conclude that the larger the bandwidth ω_b, the faster the system response. Figure 8.1 also shows that

$$0.64\ \omega_c \leq \omega_b \leq 3.2\ \omega_c.$$

So, we expect the system response to be fast if ω_c is large. However, there is a certain limitation on how large the system bandwidth can be. This will be illustrated soon with lead lag controller design examples, and also in Chapter 9 when we discuss the effects of measurement noise and plant uncertainty.

For the second-order system considered here, the phase margin, $\varphi = (\pi + \angle G(j\omega_c))$, can be computed exactly from the formula for ω_c:

$$\varphi = \frac{\pi}{2} - \tan^{-1}(\frac{\omega_c}{2\zeta\omega_o}) = \tan^{-1}(2\zeta\sqrt{\sqrt{1+4\zeta^2}+2\zeta^2}),$$

which is approximately a linear function of ζ, see Figure 8.2. Recall that as ζ increases, the percent overshoot in the step response decreases. So, large phase margin automatically implies small percent overshoot in the step response.

For large-order systems, translation of the time domain design objectives (settling time and percent overshoot) to frequency domain objectives (bandwidth and phase margin) is more complicated; but the guidelines developed for second-order systems usually extend to more general classes of systems.

We now return to our original design problem, which deals with phase margin and steady state error requirements only. Suppose that $G_1 = PC_1$ is determined; steady state error requirement is met, phase margin is to be improved by an additional controller C_2.

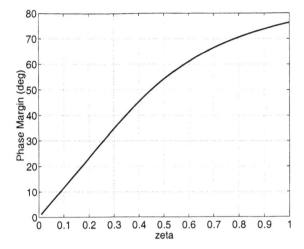

Figure 8.2: Phase margin versus ζ.

8.1 Lead Controller Design

A controller in the form

$$C_2(s) = C_{\text{lead}}(s) = \frac{(1 + \alpha \tau s)}{(1 + \tau s)} \quad \text{where} \quad \alpha > 1, \ \tau > 0,$$

is a lead controller, α and τ are to be determined.

Asymptotic Bode plots of $C_{\text{lead}}(j\omega)$ are illustrated in Figure 8.3. The phase is positive for all $\omega > 0$. Hence the phase of $G_1(j\omega)$ is increased with the addition of the lead term $C_2(j\omega)$. The largest phase increase ϕ is achieved at the frequency

$$\omega_o = \frac{1}{\tau \sqrt{\alpha}}.$$

The exact value of ϕ is given by

$$\sin(\phi) = \frac{\alpha - 1}{\alpha + 1}.$$

Note that $\phi \in (0^\circ, 90^\circ)$. See Figure 8.4 for a plot of ϕ versus α.

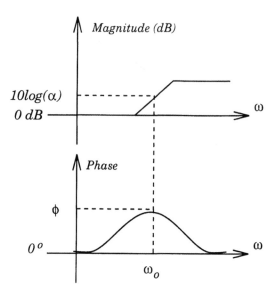

Figure 8.3: Bode plots of a lead controller.

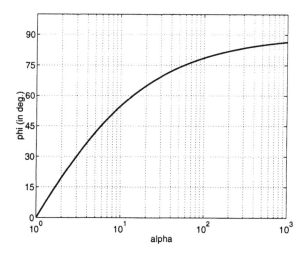

Figure 8.4: Phase lead ϕ versus α.

The basic idea behind lead controller design is to increase the phase of G_1, to increase the phase margin. Therefore, the largest phase lead ϕ should occur at the crossover frequency, i.e. w_o should be the crossover frequency for $G = G_1 C_2$. Since the magnitude of G_1 is increased by $10 \log_{10}(\alpha)$ at w_o, the crossover frequency for $G = G_1 C_2$ is the frequency at which the magnitude of G_1 is $-10 \log_{10}(\alpha)$. Once α is determined, w_o can be read from the Bode magnitude plot G_1 and then

$$\tau = \frac{1}{w_o \sqrt{\alpha}}.$$

How is α selected from the desired phase margin? To understand the effect of additional term C_2 on the phase margin, first let w_c be the crossover frequency for G_1 and let

$$\varphi_{\text{old}} = \pi + \angle G_1(jw_c) \tag{8.1}$$

be the phase margin of the system with $PC = G_1$. Then define

$$\varphi_{\text{des}} = \pi + \angle G_1(jw_o) + \angle C_2(jw_o) \tag{8.2}$$

as the desired phase margin of the system with $PC = G_1 C_2$ (in this case the crossover shifts from w_c to w_o). Since $\angle C_2(jw_o) = \phi$, equations (8.1) and (8.2) yield

$$\phi = \varphi_{\text{des}} - \varphi_{\text{old}} + (\angle G_1(jw_c) - \angle G_1(jw_o)). \tag{8.3}$$

Equation (8.3) is the basis for the lead controller design method summarized below:

Step 0. Given Bode plots of $G_1(jw)$, determine the phase margin φ_{old} and crossover frequency w_c. Desired phase margin $\varphi_{\text{des}} > \varphi_{\text{old}}$ is also given.

Step 1. Pick a small safety angle $\varphi_{\text{saf}} \approx 5°$ to $10°$ and define

$$\phi = \varphi_{\text{des}} - \varphi_{\text{old}} + \varphi_{\text{saf}}.$$

Step 2. Calculate α from the formula

$$\alpha = \frac{1 + \sin(\phi)}{1 - \sin(\phi)}.$$

Step 3. From the Bode magnitude plot of $G_1(j\omega)$ find the frequency ω_o for which

$$20\log_{10}|G_1(j\omega_o)| = -10\log_{10}(\alpha)$$

Step 4. Check if

$$\varphi_{\text{saf}} \geq (\angle G_1(j\omega_c) - \angle G_1(j\omega_o)).$$

If yes, go to Step 5. If no, go to Step 1 and increase φ_{saf} by 5° (if this leads to $\phi \geq 90°$ stop iterations: desired phase margin cannot be achieved by a first-order additional lead term).

Step 5. Define

$$\tau = \frac{1}{\omega_o\sqrt{\alpha}}.$$

Draw the Bode plots of G_1C_2, and check the phase margin.

The potential problem with the test in Step 4 is the following: As ϕ increases, α and ω_o increase. On the other hand, if the phase of $G_1(j\omega)$ decreases very fast for $\omega > \omega_c$, then the difference

$$(\angle G_1(j\omega_c) - \angle G_1(j\omega_o))$$

can be larger than φ_{saf} as ω_o increases. Therefore, lead controllers are not suitable for such systems.

Example 8.2 The system designed in Example 8.1 did not satisfy the desired phase margin of $\varphi_{\text{des}} = 45°$. The Bode plots of $G_1(s) = \frac{2}{s(s+0.1)}$, shown in Figure 8.5, give $\varphi_{\text{old}} = 4°$ and $\omega_c = 1.4$ rad/sec. By defining

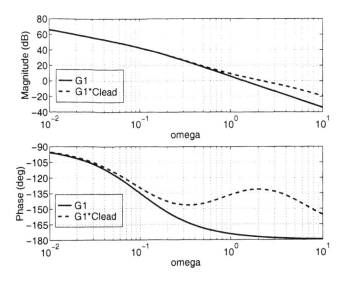

Figure 8.5: Bode plots of G_1 and $G = G_1 C_{\text{lead}}$.

$\varphi_{\text{saf}} = 5°$ additional phase is calculated as $\phi = 45° - 4° + 5° = 46°$. This leads to $\alpha = 6.13$ and $-10 \log_{10}(\alpha) = -7.87$ dB. From Figure 8.5, we see that the magnitude of $G_1(j\omega)$ drops to -7.87 dB at $\omega_o = 2.2$ rad/sec. Also note that $5°$ safety is sufficient. Indeed,

$$5° \geq (\angle G_1(j\omega_c) - \angle G_1(j\omega_o)) = (-176° - (-177.5°)) = 1.5°.$$

Example 8.3 For the system considered in the previous example, introduce a time delay in the plant:

$$G_1(s) = P(s) = \frac{2e^{-hs}}{s(s + 0.1)}$$

and study the effect of time delay on phase margin. Let $h = 0.5$ sec. Then, the Bode plots of G_1 are modified as shown in Figure 8.6. In this case, the uncompensated system is unstable, with phase margin $\varphi_{\text{old}} \cong -30°$. Try to stabilize this system by a lead controller with $\varphi_{\text{des}} = 10°$ and $\varphi_{\text{saf}} = 6°$. For these values $\phi = 46°$ (same as in

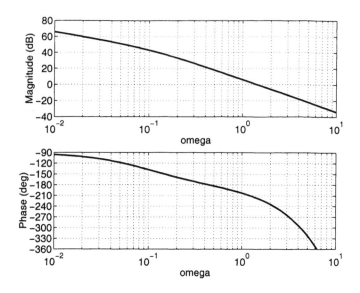

Figure 8.6: Bode plots of G_1 for Example 8.3.

Example 8.2), hence $\alpha = 6.13$ and $\omega_o = 2.2$ rad/sec. But in this case, the phase plot in Figure 8.6 shows that

$$(\angle G_1(j\omega_c) - \angle G_1(j\omega_o)) \cong (-210° - (-240°)) = 30° > \varphi_{\text{saf}}.$$

The lead controller designed this way does not even stabilize the feedback system, because

$$\angle G_1(j\omega_o) + \phi = -194° \leq -180° \,,$$

which means that the phase margin is $-14°$. The main problem here is the fact that the phase decreases very rapidly for $\omega > \omega_c = 1.4$ rad/sec. On the other hand, in the frequency region $\omega \in [0 \,, \, 0.3]$, the phase is approximately the same as the phase of G_1 without time delay. If the crossover frequency can be brought down to $\omega \approx 0.2$ to 0.3 range, then a lead controller can be designed properly. There are two ways to reduce the crossover frequency: (i) by reducing the gain K_c, or (ii) by

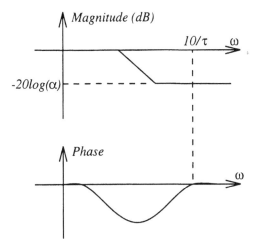

Figure 8.7: Bode plots of a lag controller.

adding a lag controller in the form

$$C_{\text{lag}}(s) = \frac{(1+\tau s)}{(1+\alpha\tau s)} \qquad \alpha > 1, \quad \tau > 0. \tag{8.4}$$

The first alternative is not acceptable, because it increases the steady state error by reducing the DC gain of G_1. Lag controllers do not change the DC gain, but they reduce the magnitude in the frequency range of interest to adjust the crossover frequency.

8.2 Lag Controller Design

As defined by (8.4) a lag controller is simply the inverse of a lead controller:

$$C_2(s) = C_{\text{lag}}(s) = \frac{(1+\tau s)}{(1+\alpha\tau s)} \quad \text{with } \alpha > 1 \text{ and } \tau > 0.$$

Typical Bode plots of $C_{\text{lag}}(j\omega)$ are shown in Figure 8.7.

The basic idea in lag controller design is to reduce the magnitude of G_1 to bring the crossover frequency to a desired location. To be more precise, suppose that with the phase lag introduced by the controller C_2, the overall phase is above $-180°$ for all $\omega < \omega_c$, and at a certain "desired" crossover frequency $\omega_d < \omega_c$ the phase satisfies

$$\varphi_{\text{des}} = \pi + \angle G_1(j\omega_d) + \angle C_2(j\omega_d).$$

Then, by choosing

$$\alpha = |G_1(j\omega_d)| > 1$$

the new crossover frequency is brought to ω_d and the desired phase margin is achieved: for this purpose we must choose $\frac{10}{\tau} \leq \omega_d$. Usually τ is chosen in such a way that

$$\omega_d = 10/\tau$$

where ω_d is determined from the phase plot of $G_1(j\omega)$ as the frequency at which

$$\angle G_1(j\omega_d) + \pi = \varphi_{\text{des}} + 5°$$

(note that $\angle C_2(j\omega_d) \approx -5°$). Thus, lag controller design procedure does not involve any trial and error type of iterations. In this case, controller parameters α and τ are determined from the Bode plots of G_1 directly.

Example 8.4 We have seen that the time delay system studied in Example 8.3 was impossible to stabilize by a first-order lead controller. Bode plots shown in Figure 8.6 illustrate that for a stable system with $\varphi_{\text{des}} = 15°$, the crossover frequency should be $\omega_d = 0.2$ rad/sec (the

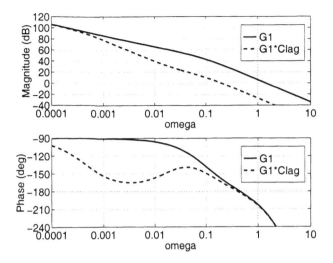

Figure 8.8: Bode plots of G_1 and $G_1 C_{\text{lag}}$ for Example 8.4.

phase of G_1 is $-160°$ at 0.2 rad/sec). The magnitude of G_1 at ω_d is
about 33 dB, that means $\alpha = 45 \equiv 33$ dB. Choosing $\tau = 10/\omega_d = 50$
the lag controller is

$$C_2(s) = C_{\text{lag}}(s) = \frac{1 + 50 \ s}{1 + 2250 \ s}.$$

The Bode plots of the lag compensated and uncompensated systems
are shown in Figure 8.8. As expected, the phase margin is about 15°.
Although 15° is not a very large phase margin, considering that the
uncompensated system had $-30°$ phase margin, this is a significant
improvement.

8.3 Lead–Lag Controller Design

Lead–lag controllers are combinations of lead and lag terms in the form

$$C_2(s) = \frac{(1 + \alpha_1 \tau_1 s)}{(1 + \tau_1 s)} \frac{(1 + \tau_2 s)}{(1 + \alpha_2 \tau_2 s)}$$

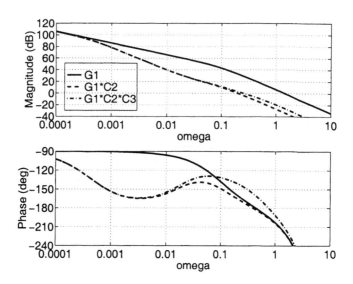

Figure 8.9: Bode plots of G_1, G_1C_2, and $G_1C_2C_3$.

where $\alpha_i > 1$ and $\tau_i > 0$ for $i = 1, 2$; and (α_1, τ_1) are determined from lead controller design principles, (α_2, τ_2) are determined from lag controller design guidelines.

Example 8.5 (Example 8.4 continued) Now, for the lag compensated system, let $C_3(s)$ be a lead controller, such that the phase margin of the lead lag compensated system $G_1C_2C_3$ is $\varphi_{\text{des}} = 30°$. This time assuming $\varphi_{\text{saf}} = 10°$, additional phase lead is $\phi = 30° - 15° + 10° = 25°$. That gives $\alpha_1 = 2.46$, and $-10\log_{10}(\alpha_1) = -4$ dB. The new crossover frequency is $\omega_o = 0.26$ rad/sec (this is the frequency at which the magnitude of G_1C_2 is -4 dB). So, $\tau_1 = 1/(0.26\sqrt{2.46}) = 2.45$ and the lead controller is

$$C_3(s) = \frac{1 + 6.027\ s}{1 + 2.45\ s}.$$

The Bode plots of the lead lag compensated system, shown in Figure 8.9, verify that the new system has $30°$ phase margin as desired.

8.4 PID Controller Design

If a controller includes three terms in parallel: proportional (P), integral (I), and derivative (D) actions, it is said to be a PID controller. Such controllers are very popular in industrial applications because of their simplicity. PID controllers can be seen as extreme cases of lead–lag controllers. This point will be illustrated shortly.

General form of a PID controller is

$$C_{\text{PID}}(s) = K_p + \frac{K_i}{s} + K_d s.$$

Let $e(t)$ be the input to the controller $C_{\text{PID}}(s)$. Then the controller output $u(t)$ is a linear combination of $e(t)$, its integral, and its derivative:

$$u(t) = K_p e(t) + K_i \int_0^t e(\tau)\mathrm{d}\tau + K_d \dot{e}(t).$$

Unless $K_d = 0$ (i.e., the derivative term is absent) the PID controller is *improper*, so it is physically impossible to realize such a controller. The difficulty is in the *exact implementation* of the derivative action. In practice, the derivative of $e(t)$ can be approximated by a proper term:

$$C_{\text{PID}}(s) \cong K_p + \frac{K_i}{s} + \frac{K_d s}{1 + \varepsilon s}$$

where ε is a small positive number.

When $K_d = 0$, we have a PI controller

$$C_{\text{PI}} = \frac{K_i(1 + \tau s)}{s} \qquad \text{where } \tau = K_p/K_i.$$

Typical Bode plots of a PI controller are shown in Figure 8.10. This figure shows that the DC gain of the open-loop transfer function is increased, hence the steady state error for step (or ramp) reference input is decreased. In fact, the system type is increased by one. By choosing $K_p < 1$ the magnitude can be decreased near the crossover frequency

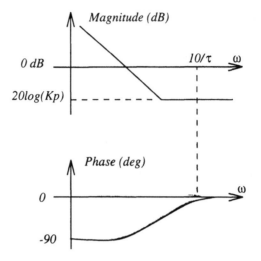

Figure 8.10: Bode plots of a PI controller.

and, by adjusting τ and K_p, desired phase margin can be achieved. This is reminiscent of the lag controller design. The only difference here is that the DC gain is higher, and the phase lag at low frequencies is larger. As an example, consider G_1 of Example 8.4: a PI controller can be designed by choosing $\tau = 50 = K_p/K_i$ and $\alpha = 45 = \frac{1}{K_p}$. This PI controller achieves $15°$ phase margin, as does the lag controller designed earlier.

Exercise:

(a) Draw the Bode plots for the system $G_1 C_{\mathrm{PI}}$, with the PI controller ($K_p = 1/45$ and $K_i = K_p/50$) determined from the lag controller design principles and verify the phase margin.

(b) Now modify the PI controller (i.e., select different values of K_p and K_i from lag controller design principles) so that the the the crossover frequency ω_d is greater or equal to 0.1 rad/sec, and the phase margin is greater or equal to $30°$.

A PD (proportional plus derivative) controller can be implemented in a proper fashion as

$$C_{\mathrm{PD}}(s) = \frac{K_p(1 + \frac{K_d}{K_p}s)}{(1 + \varepsilon s)}$$

where $\varepsilon > 0$ is a fixed small number. Usually $\varepsilon \ll K_d/K_p$. Note that by defining $\varepsilon = \tau$ and $\frac{K_d}{K_p} = \alpha\tau$, PD controller becomes a lead controller with $\alpha > 1$, $\tau > 0$ and an adjustable DC gain $C_{\mathrm{PD}}(0) = K_p$. Thus, we can design PD controllers using lead controller design principles.

By cascading a PI controller with a PD controller, we obtain a PID controller. To see this, let

$$C_{\mathrm{PD}}(s) = \frac{(K_{p1} + K_{d1}s)}{(1 + \varepsilon s)} \quad \text{and} \quad C_{\mathrm{PI}}(s) = K_{p2} + \frac{K_{i2}}{s}$$

then

$$C_{\mathrm{PID}}(s) = C_{\mathrm{PI}}(s)C_{\mathrm{PD}}(s) = \left(K_p + \frac{K_i}{s} + K_d s\right)(1 + \varepsilon s)^{-1}$$

where $K_p = (K_{p1}K_{p2} + K_{i2}K_{d1})$, $K_i = K_{i2}K_{p1}$ and $K_d = K_{d1}K_{p2}$. Therefore, PID controllers can be designed by combining lead and lag controller design principles, like lead–lag controllers.

8.5 Exercise Problems

1. Design an appropriate controller (lead, or lag, or lead–lag) for the system

$$G_1(s) = \frac{Ke^{-4s}(s^2 + 0.03s + 0.01)(s + 1)}{s^2(s^2 + 0.2s + 0.04)}$$

so that the phase margin is 30° when $K = 0.02$.

2. What is the largest phase margin achievable with a first-order lead controller for the plant

$$G_1(s) = \frac{2e^{-hs}}{s(s+0.1)}$$

(a) when $h = 0$?

(b) when $h = 0.3$?

(c) when $h = 0.5$?

Chapter 9

Principles of Loopshaping

The main goal in lead–lag controller design methods discussed earlier was to achieve good phase margin by adding extra lead and/or lag terms to the controller. During this process the magnitude and the phase of the open *loop* transfer function $G = PC$ is *shaped* in some desired fashion. So, lead–lag controller design can be considered as basic loopshaping. In this chapter, we consider tracking and noise reduction problems involving *minimum phase* plants and controllers.

9.1 Tracking and Noise Reduction Problems

Control problems studied in this chapter are related to the feedback system shown in Figure 9.1. Here $y_o(t)$ is the plant output; its noise corrupted version $y(t) = y_o(t) + n(t)$ is available for feedback control. We assume that there are no right half plane pole zero cancelations in the product $P(s)C(s) = G(s)$ (otherwise the feedback system is unstable) and that the open-loop transfer function $G(s)$ is minimum phase.

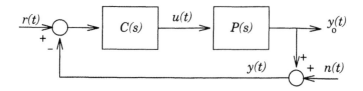

Figure 9.1: Feedback system considered for loopshaping.

A proper transfer function in the form

$$G(s) = e^{-hs}\,\frac{N_G(s)}{D_G(s)}$$

(where (N_G, D_G) is a pair of coprime polynomials) is said to be *minimum phase* if $h = 0$ and both $N_G(s)$ and $D_G(s)$ have no roots in the open right half plane. The importance of this assumption will be clear when we discuss Bode's gain–phase relationship.

The reference, $r(t)$, and measurement noise, $n(t)$, belong to

$$r \in \mathcal{R} \;=\; \{R(s) = W_r(s)R_o(s) \quad : \quad \|r_o\|_2 \le 1\}$$
$$n \in \mathcal{N} \;=\; \{N(s) = W_n(s)N_o(s) \quad : \quad \|n_o\|_2 \le 1\}$$

where W_r and W_n are filters defining these classes of signals. Typically, the reference input is a low-frequency signal; e.g., when $r(t)$ is unit step function $R(s) = 1/s$ and hence $|R(j\omega)|$ is large at low frequencies and low at high frequencies. So, $W_r(s)$ is usually a low-pass filter. The measurement noise is typically a high-frequency signal, which means that W_n is a high-pass filter. It is also possible to see W_r and W_n as reference and noise generating filters with arbitrary inputs r_o and n_o having bounded energy (normalized to unity).

Example 9.1 Let $r(t)$ be

$$r(t) = \begin{cases} t/\delta & \text{for } 0 \le t \le \delta \\ 1 & \text{for } t > \delta. \end{cases}$$

For $0 < \delta \ll 1$, this signal approximates the unit step. There exist a filter $W_r(s)$ and a finite energy input $r_o(t)$ that generate $r(t)$. To see this, let us choose $W_r(s) = K/s$, with some gain $K > 0$, and define

$$r_o(t) = \begin{cases} 1/\sqrt{\delta} & \text{for } 0 \leq t \leq \delta \\ 0 & \text{for } t > \delta. \end{cases}$$

Verify that r_o has unit energy. Also check that when $K = 1/\sqrt{\delta}$, output of the filter $W_r(s)$ with input $r_o(t)$ is indeed $r(t)$.

There are infinitely many reference signals of interest r in the set \mathcal{R}. The filter W_r emphasizes those signals r_o for which $|R_o(j\omega)|$ is large at the frequency region where $|W_r(j\omega)|$ is large, and it de-emphasizes all inputs r_o such that the magnitude $|R_o(j\omega)|$ is mainly concentrated to the region where $|W_r(j\omega)|$ is small. Similar arguments can be made for W_n and the class of noise signals \mathcal{N}.

Tracking error $e(t) = r(t) - y_o(t)$ satisfies

$$\begin{aligned} E(s) &= S(s)R(s) + T(s)N(s) \quad \text{where} \\ S(s) &= \frac{1}{1+G(s)} \quad \text{and} \quad T(s) = \frac{G(s)}{1+G(s)} = 1 - S(s) \,. \end{aligned}$$

For good tracking and noise reduction we desire

$$\begin{aligned} |S(j\omega)R(j\omega)| &\ll 1 \quad \forall \ \omega \\ |T(j\omega)N(j\omega)| &\ll 1 \quad \forall \ \omega \,, \end{aligned}$$

i.e., $|S(j\omega)|$ should be small in the frequency region where $|R(j\omega)|$ is large and $|T(j\omega)|$ should be small whenever $|N(j\omega)|$ is large. Since

$$S(s) + T(s) = 1$$

it is impossible to make both $|S(j\omega)|$ and $|T(j\omega)|$ small at the same frequency ω. Fortunately, in most practical cases $|N(j\omega)|$ is small when $|R(j\omega)|$ is large, and vice versa.

We can formally define the following performance problem: given a positive number γ_r, design $G(s)$ so that *the largest tracking error energy, due to any reference input r from the set \mathcal{R}, is less than or equal to γ_r*. A dual problem for noise reduction can be posed similarly for a given $\gamma_n > 0$. These problems are equivalent to:

$$\begin{aligned} \sup_{\substack{r \in \mathcal{R} \\ n = 0}} \quad \|e\|_2 \leq \gamma_r \quad \Longleftrightarrow \quad \|W_r S\|_\infty \leq \gamma_r \end{aligned} \tag{9.1}$$

$$\begin{aligned} \sup_{\substack{n \in \mathcal{N} \\ r = 0}} \quad \|e\|_2 \leq \gamma_n \quad \Longleftrightarrow \quad \|W_n T\|_\infty \leq \gamma_n. \end{aligned} \tag{9.2}$$

The performance objective stated via (9.1) is satisfied if and only if

$$|S(j\omega)| \leq \frac{\gamma_r}{|W_r(j\omega)|} \quad \forall \ \omega. \tag{9.3}$$

Clearly, the system has "good tracking performance" if γ_r is "small." Similarly, the effect of the noise n, on the system output y_o, is "small" if the inequality

$$|T(j\omega)| \leq \frac{\gamma_n}{|W_n(j\omega)|} \quad \forall \ \omega \tag{9.4}$$

holds with a "small" γ_n. Hence, we can see γ_r and γ_n as performance indicators: the smaller these numbers the better the performance.

Assume that W_r is a low-pass filter with magnitude

$$|W_r(j\omega)| \quad \begin{cases} \gg \gamma_r & \text{for } 0 \leq \omega \leq \omega_{\text{low}} \\ \approx 0 & \text{for } \omega > \omega_{\text{low}} \end{cases}$$

for some ω_{low} indicating the "low-frequency" region of interest. Then, (9.3) is equivalent to having

$$\frac{1}{|1 + G(j\omega)|} \ll 1 \quad \forall \ \omega \leq \omega_{\text{low}}$$

which means that $|G(j\omega)| \gg 1$ in the low-frequency band. In this case, $|S(j\omega)| \leq (|G(j\omega)| - 1)^{-1}$, and hence, by choosing

$$\frac{1}{|G(j\omega)| - 1} \leq \frac{\gamma_r}{|W_r(j\omega)|} \quad \forall \; \omega \leq \omega_{\text{low}}$$

the design condition (9.3) is automatically satisfied. In conclusion, G should be such that

$$|G(j\omega)| \geq 1 + \gamma_r^{-1}|W_r(j\omega)| \quad \forall \; \omega \leq \omega_{\text{low}}. \tag{9.5}$$

For the high-pass filter W_n, we assume

$$|W_n(j\omega)| \quad \begin{cases} \approx 0 & \text{for } 0 \leq \omega < \omega_{\text{high}} \\ \gg \gamma_n & \text{for } \omega \geq \omega_{\text{high}} \end{cases}$$

where $\omega_{\text{high}} \gg \omega_{\text{low}}$. Then, the design condition (9.4) implies

$$\frac{|G(j\omega)|}{|1 + G(j\omega)|} \ll 1 \quad \forall \; \omega \geq \omega_{\text{high}}$$

i.e., we want $|G(j\omega)| \ll 1$ in the high-frequency band. In this case, the triangle inequality gives $|T| \leq |G|(1 - |G|)^{-1}$ and we see that if

$$\frac{|G(j\omega)|}{1 - |G(j\omega)|} \leq \frac{\gamma_n}{|W_n(j\omega)|} \quad \forall \; \omega \geq \omega_{\text{high}}$$

then G satisfies (9.4). Thus, in the high-frequency band $|G(j\omega)|$ should be bounded as

$$|G(j\omega)| \leq (1 + \gamma_n^{-1}|W_n(j\omega)|)^{-1} \quad \forall \; \omega \geq \omega_{\text{high}}. \tag{9.6}$$

In summary, low and high-frequency behaviors of $|G|$ are characterized by (9.5) and (9.6). The magnitude of G should be much larger than 1 at low frequencies and much smaller than 1 at high frequencies.

Therefore, the gain crossover frequencies are located in the transition band (ω_{low}, ω_{high}). In the next section, we will see a condition on the magnitude of G around the crossover frequencies.

9.2 Bode's Gain–Phase Relationship

There is another key design objective in loopshaping: the phase margin should be as large as possible. Now we will try to understand the implications of this design goal on the magnitude of G. Recall that the phase margin is

$$\varphi = \min\{\pi + \angle G(j\omega_{ci}) \quad : \quad i = 1, \ldots, \ell\}$$

where ω_{ci}'s are the crossover frequencies, i.e., $|G(j\omega_{ci})| = 1$. Without loss of generality assume that $\ell = 1$, in other words gain crossover occurs at a single frequency ω_c (since $G = PC$ is designed here, we can force G to have single crossover). For good phase margin $\angle G(j\omega_c)$ should be significantly larger than $-\pi$. For example, for 60° to 90° phase margin the phase $\angle G(j\omega_c)$ is desired to be $-120°$ to $-90°$.

It is a well known mathematical fact that if $G(s)$ is a minimum phase transfer function, then its phase function $\angle G(j\omega)$ is uniquely determined from the magnitude function $|G(j\omega)|$, once the sign of its DC gain $G(0)$ is known (i.e., we should know whether $G(0) > 0$ or $G(0) < 0$). In particular $\angle G(j\omega_c)$, hence, the phase margin can be estimated from the asymptotic Bode magnitude plot. Assume $G(0) > 0$, then for any $\omega_o > 0$ the exact formula for the phase is (see [10, p. 313])

$$\angle G(j\omega_o) = \frac{1}{\pi} \int_{-\infty}^{\infty} M(\nu)W(\nu)d\nu$$

where $\nu = \ln(\omega/\omega_o)$ is the normalized frequency appearing as the integration variable (note that $\omega = \omega_o e^\nu$) and

$$M(\nu) \quad = \quad \frac{d}{d\nu}|G(j\omega_o e^\nu)|$$

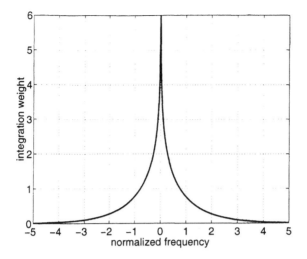

Figure 9.2: $W(\nu)$ versus ν.

$$W(\nu) \;\;=\;\; \ln(\coth(\frac{|\nu|}{2})).$$

The phase at ω_o depends on the magnitude at all frequencies $\omega \in$ $(0 \, , \, \infty)$. So, it is not an easy task to derive the whole phase function from the magnitude plot. But we are mainly interested in the phase at the crossover frequency $\omega_o = \omega_c$. The connection between asymptotic Bode plots and the above phase formula is this: $M(\nu)$ is the slope of the log–log magnitude plot of G at the normalized frequency ν. The weighting function $W(\nu)$ is independent of G. As shown in Figure 9.2, its magnitude is very large near $\nu = 0$ (in fact $W(0) = \infty$) and it decays to zero exponentially. For example, when $|\nu| > 4$ we can treat $W(\nu)$ as negligibly small: $W(\pm 4) < 0.04$.

Now assume that the slope of the log–log Bode magnitude plot is almost constant in the frequency band $\omega \in [\omega_c e^{-4} \, , \, \omega_c e^4]$, say $-n \times 20$ dB per decade. Then $M(\nu) = -n$ in the normalized frequency band

$|\nu| \leq 4$ and hence,

$$\angle G(j\omega_c) \cong -\frac{n}{\pi} \int_{-4}^{4} W(\nu)\mathrm{d}\nu \cong -\frac{n}{\pi} \int_{-\infty}^{\infty} W(\nu)\mathrm{d}\nu = -\frac{n\pi}{2}$$

(the error in the approximation of the integral of $W(\nu)$ is less than 2 percent of the exact value $\pi^2/2$). Since Bode's work, [10], the approximation $\angle G(j\omega_c) \cong -\frac{n\pi}{2}$ has been used as an important guideline for loopshaping: if the Bode magnitude plot of G decreases smoothly with a slope of -20 dB/dec, near $\omega = \omega_c$, then the phase margin will be large (close to 90°). From the above phase approximations it is also seen that if the Bode magnitude of G decreases with a slope -40 dB/dec, or faster, near $\omega = \omega_c$, then the phase margin will be small, if not negative.

In summary, we want to shape the Bode magnitude of G so that around the crossover frequency its slope is almost constant at -20 dB/dec. This assures good phase margin.

9.3 Design Example

We now have a clear picture of the desired shape of the magnitude plot for $G = PC$. Loopshaping conditions derived in the previous sections are illustrated in Figure 9.3, where G_{low} and G_{high} represent the lower and upper bounds given in (9.5) and (9.6) respectively, and the dashed lines indicate the desired slope of $|G|$ near the crossover frequency ω_c.

It is clear that the separation between ω_{low} and ω_{high} should be at least two decades, so that ω_c can be placed approximately one decade away from low and high frequencies, and -20 dB/dec slope can be assigned to $|G|$ around the crossover frequency. Otherwise, the transition frequency band will be short: in this case, if G_{low} is too large and G_{high} is too small, it will be difficult to achieve a smooth -20 dB/dec slope around a sufficiently large neighborhood of ω_c.

Another important point in the problem setup is this: since $C =$

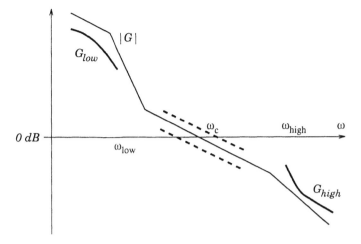

Figure 9.3: Loopshaping conditions.

G/P must be a proper function, $|G(j\omega)|$ should decay to zero at least as fast as $|P(j\omega)|$ does, for $\omega \to \infty$. Therefore, $G_{high}(\omega)$ should roll off with a slope $-\ell \times 20$ dB/dec, where ℓ is greater or equal to the relative degree of $P = N_P/D_P$ i.e.,

$$\ell \geq \deg(D_P(s)) - \deg(N_P(s)).$$

This is guaranteed if $W_n(s)$ is improper with relative degree $-\ell$.

The example we study here involves an uncertain plant

$$P(s) = P_o(s) + \Delta_P(s) \quad \text{where} \quad P_o(s) = \frac{\omega_o^2}{s^2 + 2\zeta\omega_o s + \omega_o^2}$$

$\Delta_P(s)$ is an unknown stable transfer function, $\omega_o > 0$ and $\zeta \in (0 , 1)$. In this case, the standard feedback system with uncertain P, depicted in Figure 9.4, is equivalent to the feedback system shown in Figure 9.5. The feedback signal $y(t)$ is the sum of two signals: $y_o(t)$ and $y_\delta(t)$. The latter signal is unknown due to plant uncertainty Δ_P. Using frequency

Figure 9.4: Standard feedback system with uncertain plant.

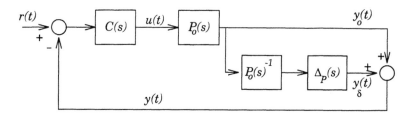

Figure 9.5: Equivalent feedback system.

domain representations,

$$Y_\delta(s) = \frac{1}{P_o(s)} \, \Delta_P(s) Y_o(s).$$

Now defining

$$k_n(s) N_o(s) := \Delta_P(s) Y_o(s) \quad \text{and} \quad W_n(s) = \frac{k_n(s)}{P_o(s)}$$

we see that Figure 9.5 is equivalent to Figure 9.1 with $P = P_o$ and $N = W_n N_o$.

Assuming that the feedback system is stable, y_o is a bounded function; that implies n_o is bounded because Δ_P is stable. The scaling factor $k_n(s)$ is introduced to normalize N_o. Large values of $|k_n(j\omega)|$ imply a large uncertainty magnitude $|\Delta_P(j\omega)|$. Conversely, if $|k_n(j\omega)|$ is small, then the contribution of the uncertainty is small at that frequency. In summary, the problem is put in the framework of Figure 9.1 by taking out the uncertainty and replacing $y_\delta(t)$ with $n(t)$. The noise $n(t)$ is from the set \mathcal{N}, which is characterized by the filter $W_n(s) = k_n(s)/P_o(s)$, where $|k_n(j\omega)|$ represents the "size" of the uncertainty.

Let us assume that $|k_n(j\omega)| \approx 0$ for $\omega < 10 \, \omega_o =: \omega_{high}$, i.e., the nominal model $P_o(j\omega)$ represents the actual plant $P(j\omega)$ very well in the low-frequency band. Note that, $|P_o(j\omega)| \ll 1$ for $\omega \geq \omega_{high}$. So we can assume that in the high-frequency region the magnitude of the uncertainty can be relatively large, e.g.,

$$k_n(j\omega) = 0.05 \, \frac{j0.1 \, \omega/\omega_o}{1 + j0.01 \, \omega/\omega_o} \qquad \forall \; \omega \geq 10 \, \omega_o.$$

In this numerical example the uncertainty magnitude is 0.05 near $\omega = 10 \, \omega_o$, and it is about 0.5 for $\omega > 100 \, \omega_o$. To normalize the frequency band, set $\omega_o = 1$. Now the upper bound (determined in (9.6))

$$G_{high}(\omega) := \frac{\gamma_n}{\gamma_n + |W_n(j\omega)|} = \frac{\gamma_n |P_o(j\omega)|}{\gamma_n |P_o(j\omega)| + |k_n(j\omega)|},$$

can be computed for any given performance indicator γ_n. Here we take $\gamma_n = 0.25$ arbitrarily.

For the tracking problem, suppose that the steady state error, e_{ss}, must satisfy

- $e_{ss} \leq \frac{1}{2}$ for unit ramp reference input,

- $|e_{ss}| \leq \frac{1}{40}$ for all sinusoidal inputs whose periods are 90 seconds or larger.

The first condition implies that

$$\lim_{s \to 0} sG(s) \geq \lim_{s \to 0} s \, \frac{2}{s} = 2,$$

thus we have a lower bound on the DC behavior of $G(s)$. Recall that for a sinusoidal input of period τ the amplitude of the steady state error is

$$|e_{ss}| = |S(j\frac{2\pi}{\tau})| \, ,$$

and this quantity should be less than or equal to $\frac{1}{40}$ for $\tau \geq 90$ sec. Therefore, by using the notation of Section 9.1, we define

$$\gamma_r^{-1}|W_r(j\omega)| = 40 \quad \forall \ \omega \in (0 \ , \ \omega_{\text{low}}]$$

where $\omega_{\text{low}} = 2\pi/90 \cong 0.07$ rad/sec. In conclusion, a lower bound for $|G(j\omega)|$ can be defined in the low-frequency region as

$$|G(j\omega)| \geq \max\{ \ \frac{2}{\omega} \ , \ (1 + 40) \ \} =: G_{low}(\omega).$$

The low-frequency lower bound, $G_{low}(\omega)$, and the high-frequency upper bound $G_{high}(\omega)$ are shown as dashed lines in Figure 9.6. The desired open-loop shape

$$G(s) = \frac{2.5(1 + 1.2s/0.125 + s^2/0.125^2)}{s(1 + 0.7s/0.07 + s^2/0.07^2)(1 + s/7)^2} \tag{9.7}$$

is designed from following guidelines:

- as $\omega \to 0$, we should have $|G(j\omega)| > 2/\omega$; the term $\frac{2.5}{s}$ takes care of this requirement;

- ω_c should be placed near 1 rad/sec, say between 0.6 rad/sec and 2 rad/sec, and the roll-off around a large neighborhood of ω_c should be -20 dB/dec;

- to have ω_c in this frequency band we need fast roll-off near $\omega \approx$ 0.07 rad/sec; the second-order term in the denominator takes care of that, its damping coefficient is chosen relatively small so that the magnitude stays above $20 \log(41)$ dB in the neighborhood of $\omega = 0.07$ rad/sec,

- for $\omega > \omega_{\text{low}}$ the slope should be constant at -20 dB/dec, so the zeros near $\omega = 0.125$ rad/sec are introduced to cancel the effect of complex poles near $\omega = 0.07$ rad/sec,

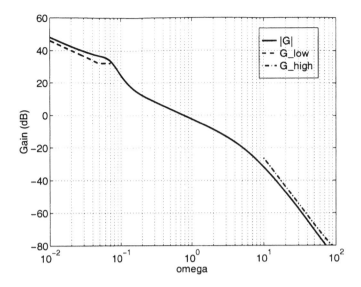

Figure 9.6: $G_{low}(\omega)$, $G_{high}(\omega)$ and $|G(j\omega)|$.

- in the frequency range $\omega_c < \omega < 10$ rad/sec the magnitude should start to roll off fast enough to stay below the upper bound, so we use double poles at $s = 7$.

For this example, verify that the gain margin is 25 dB and the phase margin is 70°. It is possible to tweak the numbers appearing in (9.7) and improve the gain and/or phase margin without violating upper and lower bound conditions. In general, loopshaping involves several steps of trial and error to obtain a "reasonably good" shape for $|G(j\omega)|$. The "best" loop shape is difficult to define.

Once $G(s)$ is determined, the controller is obtained from

$$C(s) = \frac{G(s)}{P(s)}.$$

In the above example $P(s) = P_o(s) = \frac{\omega_o^2}{s^2+2\zeta\omega_o s+\omega_o^2}$; so the controller is

$$C(s) = G(s)\frac{(s^2 + 2\zeta\omega_o s + \omega_o^2)}{\omega_o^2}, \tag{9.8}$$

where $G(s)$ is given by (9.7). The last term in (9.8) cancels the complex conjugate poles of the plant. As we have seen in Chapter 5, it is dangerous to cancel lightly damped poles of the plant (i.e., poles close to Im-axis). If the exact location of these poles are unknown, then we must be cautious. By picturing how the root locus behaves in this case, we see that the zeros should be placed to the left of the poles of the plant. An undesirable situation might occur if we overestimate ω_o and underestimate ζ in selecting the zeros of the controller: in this case, two branches of the root locus may escape to the right half plane as seen in Chapter 5. Therefore, we should use the lowest estimated value for ω_o, and highest estimated value for ζ, in $C(s)$ so that the associated branches of the root locus bend towards left.

9.4 Exercise Problems

1. For the above example let $G(s)$ be given by (9.7) and define $\zeta = 0.02$, $\omega_o = 1$, $\widehat{\zeta} := \zeta + \delta_\zeta$, $\widehat{\omega}_o = \omega_o + \delta_{\omega_o}$, where $|\delta_\zeta| \leq 0.004$ and $|\delta_{\omega_o}| \leq 0.1$.

 (a) Draw Bode plots of

 $$\widehat{G}(s) = G(s)\frac{\omega_o\,(s^2 + 2\widehat{\zeta}\widehat{\omega}_o s + \widehat{\omega}_o^2)}{\widehat{\omega}_o^2\,(s^2 + 2\zeta\omega_o s + \omega_o^2)}.$$

 for different values of δ_ζ and δ_{ω_o}, illustrating the changes in gain and phase margins.

 (b) Let $\delta_\zeta = 0.004$ and $\delta_{\omega_o} = -0.1$, and draw the root locus.

 (c) Repeat part (b) with $\delta_\zeta = -0.004$ and $\delta_{\omega_o} = +0.1$.

2. In this problem, we will design a controller for the plant

$$P(s) = \frac{(1 + 2\zeta_n s/\omega_n + (s/\omega_n)^2)}{s \left(1 + 2\zeta_d s/\omega_d + (s/\omega_d)^2\right)}$$

by using loopshaping techniques with the following weights

$$\gamma_r^{-1}|W_r(j\omega)| = \frac{k}{\omega^2} \qquad \omega_{\text{low}} = 0.1 \text{ rad/sec}$$

$$\gamma_n^{-1}|W_n(j\omega)| = \frac{0.01 \, k \, \omega^3}{\sqrt{1 + 0.001 \, \omega^2}} \qquad \omega_{\text{high}} = 10 \text{ rad/sec},$$

where $k > 0$ is a design parameter to be maximized.

(a) Find a reasonably good desired loopshape $G(s)$ for (i) $k = 0.1$, (ii) $k = 1$, and (iii) $k = 5$. What happens to the upper and lower bounds (G_{high} and G_{low}) as k increases? Does the problem become more difficult or easier as k increases?

(b) Assume that $\zeta_n = 0.3 \pm 5\%$, $\zeta_d = 0.03 \pm 8\%$, $\omega_n = 4 \pm 5\%$, $\omega_d = 2 \pm 8\%$ and let $k = 1$. Derive a controller from the result of part (a). For this system, draw the Bode plots with several different values of the uncertain plant parameters and calculate the gain, phase, and vector margins.

3. Let $P(s) = \frac{1}{(s+0.1)(s+0.01)}$. Design a controller $C(s)$ such that

(i) the system is stable with gain margin ≥ 40 dB and phase margin $\geq 50°$,

(ii) the steady state error for a unit step input is less than or equal to 10^{-4},

(iii) and the loop shaping conditions

$$\|W_r S\|_\infty \leq 1$$
$$\|W_n T\|_\infty \leq 1$$

hold for $W_r(s) = 10(s^3 + 2s^2 + 2s + 1)^{-1}$ and $W_n(s) = 0.008 \, s^2$.

Chapter 10

Robust Stability and Performance

10.1 Modeling Issues Revisited

In this chapter, we consider the standard feedback control system shown in Figure 10.1, where C is the controller and P is the plant.

As we have seen in Section 2.4, there are several approximations involved in obtaining a mathematical model of the plant. One of the reasons we desire a simple transfer function (with a few poles and zeros) for

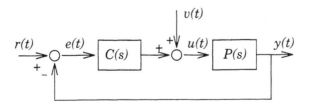

Figure 10.1: Standard feedback system.

the plant is that it is difficult to design controllers for complicated plant models. The approximations and simplifications in the plant dynamics lead to a nominal plant model. However, if the controller designed for the nominal model does not take into account the approximation errors (called plant uncertainties), then the feedback system may become unstable when this controller is used for the actual plant. In order to avoid this situation, we should determine a bound on the uncertainty and use this information when we design the controller. There are mainly two types of uncertainty: unmodeled dynamics and parametric uncertainty. Earlier in Chapter 4 we saw robust stability tests for systems with parametric uncertainty (recall Kharitanov's test, and its extensions). In this chapter, \mathcal{H}^∞-based controller design techniques are introduced for systems with dynamic uncertainty. First, in this section we review modeling issues discussed in Section 2.4, give additional examples of unmodeled dynamics, and illustrate again how to transform a parametric uncertainty into a dynamic uncertainty.

10.1.1 Unmodeled Dynamics

A practical example is given here to illustrate why we have to deal with unmodeled dynamics and how we can estimate uncertainty bounds in this case. Consider a flexible beam attached to a rigid body (e.g., a large antenna attached to a space satellite). The input is the torque applied to the beam by the rigid body and the output is the displacement at the other end of the beam. Assume that the magnitude of the torque is small, so that the displacements are small. Then, we can use a linearized model, and the input/output behavior of this system is given by the transfer function

$$P(s) = P_\infty(s) = \frac{K_0}{s^2} + \sum_{k=1}^{\infty} \frac{\alpha_k \omega_k^2}{s^2 + 2\zeta_k \omega_k s + \omega_k^2}$$

where ω_k's represent the frequency of natural oscillations (modes of the system), ζ_k's are the corresponding damping coefficients for each mode and α_k's are the weights of each mode. Typically, $\varepsilon < \zeta_k < 1$ for some $\varepsilon > 0$, $\alpha_k \to 0$ and $\omega_k \to \infty$ as $k \to \infty$. Obviously, it is not practical to work with infinitely many modes. When α_k's converge to zero very fast, we can truncate the higher-order modes and obtain a finite dimensional model for the plant. To illustrate this point, let us define the approximate plant transfer function as

$$P_N(s) = \frac{K_0}{s^2} + \sum_{k=1}^{N} \frac{\alpha_k \omega_k^2}{s^2 + 2\zeta_k \omega_k s + \omega_k^2}.$$

The approximation error between the "true" plant transfer function and the approximate model is

$$\Delta_{P_N}(s) = P_\infty(s) - P_N(s) = \sum_{k=N+1}^{\infty} \frac{\alpha_k \omega_k^2}{s^2 + 2\zeta_k \omega_k s + \omega_k^2}.$$

As long as $|\Delta_{P_N}(j\omega)|$ is "sufficiently small" we can "safely" work with the approximate model. What we mean by "safely" and how small is sufficiently small will be discussed shortly. First try to determine the peak value of the approximation error: assume $0 < \varepsilon < \zeta_k < \frac{1}{\sqrt{2}}$ for all $k \geq N$, then

$$|\Delta_{P_N}(j\omega)| \leq \sum_{k=N+1}^{\infty} \left| \frac{\alpha_k \omega_k^2}{\omega_k^2 - \omega^2 + j2\zeta_k \omega_k \omega} \right| \leq \sum_{k=N+1}^{\infty} \frac{|\alpha_k|}{\zeta_k \sqrt{2}}$$

Now suppose $|\alpha_k| \leq \zeta_k \sqrt{2} k^{-2}$ for all $k \geq N$. Then, we have the following bound for the *worst approximation error*:

$$\sup_\omega |\Delta_{P_N}(j\omega)| \leq \sum_{k=N+1}^{\infty} k^{-2} \leq \int_{x=N}^{\infty} x^{-2} dx = \frac{1}{N}.$$

For example, if we take the first 10 modes only, then the worst approximation error is bounded by $\frac{1}{10}$. Clearly, as N increases, the worst

approximation error decreases. On the other hand, we do not want to take too many modes because this complicates the model. Also note that if $|\alpha_k|$ decays to zero faster, then the bound is smaller.

In conclusion, for a plant $P(s)$, a low-order model denoted by $P_\theta(s)$ can be determined. Here θ represents the parameters of the low-order model, e.g., coefficients of the transfer function. Moreover, an upper bound of the worst approximation error can be obtained in this process:

$$|P(j\omega) - P_\theta(j\omega)| < \delta_1(\omega).$$

The pair $(P_\theta(s), \delta_1(\omega))$ captures the uncertain plant $P(s)$. The parameters represented by θ may be uncertain too. But we assume that there are known bounds on these parameters. This gives a nominal plant model $P_o(s) := P_{\theta_o}(s)$ where θ_o represent the nominal value of the parameter vector θ. See next section for a specific example.

10.1.2 Parametric Uncertainty

Now we transform a parametric uncertainty into a dynamic uncertainty and determine an upper bound on the worst approximation error. As an example, consider the above flexible beam model and take

$$P_\theta(s) = \frac{K_0}{s^2} + \frac{\alpha_1 \omega_1^2}{s^2 + 2\zeta_1 \omega_1 s + \omega_1^2}.$$

Here $\theta = [K_0, \alpha_1, \zeta_1, \omega_1]$. Suppose that an upper bound $\delta_1(\omega)$ is determined as above for the unmodeled uncertainty using bounds on α_k's and ζ_k's, for $k \geq 2$. In practice, θ is determined from the physical parameters of the beam (such as inertia and elasticity). If these parameters are not known *exactly*, then we have parametric uncertainty as well. At this point, we can also assume that the parameters of $P_\theta(s)$ are determined from system identification and parameter estimation algorithms. These techniques give upper and lower bounds, and nominal

values of the uncertain parameters. For our example, let us assume that $K_0 = 2$ and $\alpha_1 = 0.2$, are known exactly and $\zeta_1 \in [0.1 \ , \ 0.2]$ with nominal value $\zeta_1 = 0.15$, $\omega_1 \in [9 \ , \ 12]$ with nominal value $\omega_1 = 10$. Define

$$\theta_o = [2 \ , \ 0.2 \ , \ 0.15 \ , \ 10]$$

and the nominal plant transfer function

$$P_o(s) = P_{\theta_o}(s) = \frac{2}{s^2} + \frac{20}{s^2 + 3s + 100} \ .$$

The parametric uncertainty bound is denoted by $\delta_2(\omega)$:

$$|P_\theta(j\omega) - P_o(j\omega)| < \delta_2(\omega) \quad \forall \ \omega$$

for all feasible θ. Figure 10.2 shows the plot of $|P_\theta(j\omega) - P_o(j\omega)|$ for several different values of ζ_1 and ω_1 in the intervals given above. The upper bound shown here is $\delta_2(\omega) = |W_p(j\omega)|$:

$$W_p(s) = \frac{10(s + 0.15)}{(s^2 + 13s + 64)} \ .$$

Parametric uncertainty bound and the bound for unmodeled dynamic uncertainty can be combined to get an upper bound on the overall plant uncertainty

$$\Delta_P(s) = P(s) - P_o(s) \quad \text{where} \quad |\Delta_P(j\omega)| < |\delta_1(\omega)| + |\delta_2(\omega)|$$

As in Section 2.4, let $W(s)$ be an overall upper bound function such that

$$|W(j\omega)| > |\delta_1(\omega)| + |\delta_2(j\omega)| > |\Delta_P(j\omega)| \ .$$

Then, we will treat the set

$$\mathcal{P}_\Delta \quad = \quad \{P = P_o + \Delta_P(s) \ : \ |\Delta_P(j\omega)| < |W_a(j\omega)| \ \text{and}$$
$$P \text{ and } P_o \text{ have the same number of poles in } \overline{\mathbb{C}}_+\}$$

as the *set of all possible plants*.

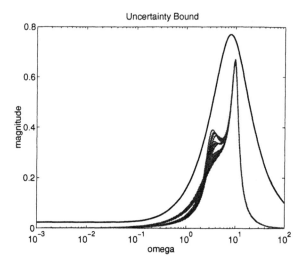

Figure 10.2: Parametric uncertainty bound.

10.2 Stability Robustness

10.2.1 A Test for Robust Stability

Given the nominal plant $P_o(s)$ and additive uncertainty bound $W(s)$, characterizing the set of all possible plants \mathcal{P}_Δ, we want to derive conditions under which a fixed controller $C(s)$ stabilizes the feedback system for all plants $P \in \mathcal{P}_\Delta$. Since $P_o \in \mathcal{P}_\Delta$, the controller C should stabilize the nominal feedback system (C, P_o). This is a necessary condition for robust stability. Accordingly, assume that (C, P_o) is stable. Then, we compare the Nyquist plot of the nominal system $G_o(s) = C(s)P_o(s)$, and the Nyquist plots of all possible systems of the form $G(s) = C(s)P(s)$, where $P \in \mathcal{P}_\Delta$. By the assumption that P_o and P have the same number of poles in $\overline{\mathbb{C}}_+$, for robust stability $G(j\omega)$ should encircle -1 as many times as $G_o(j\omega)$ does. Note that for each ω, we have

$$G(j\omega) = G_o(j\omega) + \Delta_P(j\omega)C(j\omega).$$

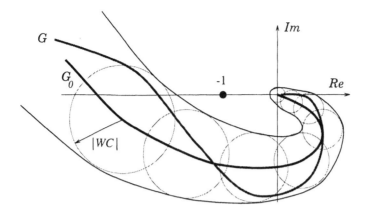

Figure 10.3: Robust stability test via Nyquist plot.

This identity implies that $G(j\omega)$ is in a circle whose center is $G_o(j\omega)$ and radius is $|W(j\omega)C(j\omega)|$. Hence, the Nyquist plot for G lies within a tube around the nominal Nyquist plot $G_o(j\omega)$. The radius of the tube is changing with ω, as shown in Figure 10.3.

By the above arguments, the system is robustly stable if and only if -1 is outside this tube, in other words,

$$|G_o(j\omega) + \Delta_P(j\omega)C(j\omega) - (-1)| > 0$$

for all ω and for all admissible $\Delta_P(j\omega)$. Since the magnitude of $\Delta_P(j\omega)$ can be as large as $(|W(j\omega)| - \epsilon)$ where $\epsilon \searrow 0$ and its phase is not restricted, we have robust stability if and only if

$$|1 + G_o(j\omega)| - |W(j\omega)C(j\omega)| \geq 0 \quad \text{for all } \omega,$$

which is equivalent to

$$|W(j\omega)C(j\omega)(1 + P_o(j\omega)C(j\omega))^{-1}| \leq 1 \quad \text{for all } \omega,$$

i.e.

$$\boxed{\|WCS\|_\infty \le 1}$$ (10.1)

where $S = (1 + P_oC)^{-1}$ is the nominal sensitivity function.

It is very important to understand the difference between the set of plants \mathcal{P}_Δ, (where $\Delta_P(s)$ can be *any* transfer function such that P and P_o have the same number of right half plane poles and $|\Delta_P(j\omega) < |W(j\omega)|)$ and a set of uncertain plants described by the parametric uncertainty. For example, consider the uncertain time delay example of Section 2.4.2

$$P(s) = \frac{e^{-hs}}{(s + 1)} \qquad h \in [0\ ,\ 0.2]$$

with $P_o(s) = \frac{1}{(s+1)}$ and $W(s) = \frac{0.0025\ (100\ s+1)}{(1.2\ s+1)(0.06\ s+1)}$. It was noted that the set

$$\mathcal{P}_h := \{P(s) = \frac{e^{-hs}}{(s + 1)} \quad : \quad h \in [0\ ,\ 0.2]\}$$

is a proper subset of

$$\mathcal{P}_\Delta := \{P_o + \Delta\ :\ \Delta(s) \text{ is stable, } |\Delta(j\omega)| < |W(j\omega)|\ \forall\ \omega\}.$$

Hence, if we can prove robust stability for all plants in \mathcal{P}_Δ, then we have robust stability for all plants in \mathcal{P}_h. The converse is not true. In fact, we have already determined a necessary and sufficient condition for robust stability for all plants in \mathcal{P}_h: for a fixed controller C, from the Bode plots of the nominal system $G_o = P_oC$ determine the phase margin, φ in radians, and crossover frequency, ω_c in radians per second, then the largest time delay the system can tolerate is $h_{\max} = \varphi/\omega_c$. That is, the closed-loop system is robustly stable for plants in \mathcal{P}_h if and only if $0.2 < h_{\max}$. The controller may be such that this condition

holds, but (10.1) does not. So, there is a certain degree of conservatism in transforming a parametric uncertainty into a dynamic uncertainty. This should be kept in mind.

Now we go back to robust stability for the class of plants \mathcal{P}_Δ. If a multiplicative uncertainty bound $W_m(s)$ is given instead of the additive uncertainty bound $W(s)$, we use the following relationship:

$$P = P_o + \Delta_P = P_o(1 + P_o^{-1}\Delta_P)$$

i.e., multiplicative uncertainty bound satisfies

$$|P_o^{-1}(j\omega)\Delta_P(j\omega)| < |W_m(j\omega)| \quad \text{for all } \omega.$$

Then, in terms of W_m, the condition (10.1) can be rewritten as

$$\|W_m P_o C (1 + P_o C)^{-1}\|_\infty \le 1. \tag{10.2}$$

Recall that the sensitivity function of the nominal feedback system (C, P_o) is defined as $S = (1 + P_o C)^{-1}$ and the complementary sensitivity is $T = 1 - S = P_o C(1 + P_o C)^{-1}$. So, the feedback system (C, P) is stable for all $P \in \mathcal{P}_\Delta$ if and only if (C, P_o) is stable and

$$\boxed{\|W_m T\|_\infty \le 1.} \tag{10.3}$$

For a given controller C stabilizing the nominal feedback system, it is possible to determine the amount of plant uncertainty that the system can tolerate. To see this, rewrite the inequality (10.3) as

$$|W_m(j\omega)| \le |T(j\omega)|^{-1} \quad \text{for all } \omega.$$

Hence, the largest multiplicative plant uncertainty that the system can tolerate at each ω is $|T(j\omega)|^{-1}$.

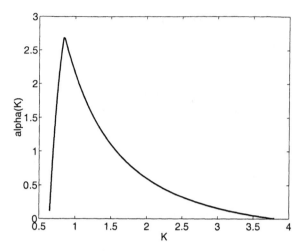

Figure 10.4: α versus K.

Example 10.1 Let

$$P_o(s) \;=\; \frac{10(s + 0.2)}{s(s - 0.5)(s^2 + 4s + 12)}$$

$$W(s) \;=\; \alpha\widehat{W}(s) = \alpha\frac{0.4(s + 0.25)}{(s + 2)}.$$

The system has "good" robustness for this class of plants if α is "large." We are now interested in designing a proportional controller $C(s) = K$ such that $\|\alpha\widehat{W}CS\|_\infty \leq 1$ for the largest possible α. First, we need to find the range of K for which (C, P_o) is stable. By the Routh-Hurwitz test, we find that K must be in the interval $(0.65 , 3.8)$. Then, we define

$$\alpha^{-1}(K) = \sup_\omega \left| \frac{\widehat{W}(j\omega)K}{1 + KP_o(j\omega)} \right|.$$

By plotting the function $\alpha(K)$ versus $K \in (0.65 , 3.8)$, see Figure 10.4, we find the largest possible α and the corresponding K: $\alpha_{\max} = 2.72$ is achieved at $K = 0.845$.

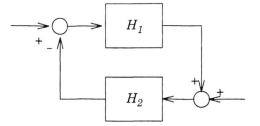

Figure 10.5: Feedback system with small gain $\|H_1 H_2\|_\infty < 1$.

10.2.2 Special Case: Stable Plants

Suppose that the nominal plant, $P_o(s)$, and the additive uncertainty, $\Delta(s)$, are stable. Then, we can determine a *sufficient* condition for robust stability using *the small gain theorem* stated below. This theorem is very useful in proving robust stability of uncertain systems. Here it is stated for linear systems, but the same idea can be extended to a large class of nonlinear systems as well.

Theorem 10.1 (Small Gain) *Consider the feedback system shown in Figure 10.5, where H_1 and H_2 are stable linear systems. If*

$$|H_1(j\omega)H_2(j\omega)| < 1 \quad \text{for all } \omega, \tag{10.4}$$

then the feedback system is stable.

The proof of this theorem is trivial: by (10.4) the Nyquist plot of $G = H_1 H_2$ remains within the unit circle, so it cannot encircle the critical point. By the Nyquist stability criterion, the feedback system is stable.

Now consider the standard feedback system (C, P). Suppose that the nominal feedback system (C, P_o) is stable. Then the system is robustly stable if and only if $S_\Delta = (1 + (P_o + \Delta_P)C)^{-1}$ and CS_Δ are stable. It

is a simple exercise to show that

$$S_\Delta = S(1 + \Delta_P CS)^{-1}.$$

Since (C, P_o) is stable, S, CS and P_oS are stable transfer functions. Now, applying the small gain theorem, we see that if

$$|\Delta_P(j\omega)C(j\omega)S(j\omega)| < |W(j\omega)C(j\omega)S(j\omega)| \le 1 \quad \text{for all} \quad \omega,$$

then S_Δ, CS_Δ and PS_Δ are stable for all $P \in \mathcal{P}_\Delta$, which means that the closed-loop system is robustly stable.

10.3 Robust Performance

Besides stability, we are also interested in the performance of the closed-loop system. In this section, performance will be measured in terms of *worst tracking error energy*. As in Chapter 9, define the set of all possible reference inputs

$$\mathcal{R} = \{R(s) = W_r(s)R_o(s) \quad : \quad \|r_o\|_2 \le 1\}.$$

Let $e(t) := r(t) - y(t)$ be the tracking error in the standard unity feedback system formed by the controller C and the plant P. Recall that the worst error energy (over all possible r in \mathcal{R}), is

$$\sup_{r \in \mathcal{R}} \|e\|_2 = \sup_{\omega} |W_r(j\omega)(1 + P(j\omega)C(j\omega))^{-1}| = \|W_r(1 + PC)^{-1}\|_\infty.$$

If a desired error energy bound, say γ_r, is given, then the controller should be designed to achieve

$$\|W_r(1 + PC)^{-1}\|_\infty \le \gamma_r \tag{10.5}$$

for all possible $P = P_o + \Delta_P \in \mathcal{P}_\Delta$. This is the *robust performance* condition. Of course, the definition makes sense only if the system is

robustly stable. Therefore, we say that the controller achieves robust performance if it is a robustly stabilizing controller and (10.5) holds for all $P \in \mathcal{P}_\Delta$. It is easy to see that this inequality holds if and only if

$$|W_1(j\omega)S(j\omega)| < |1 + \Delta_P(j\omega)C(j\omega)S(j\omega)|$$

for all ω and all admissible Δ_P, where $S(s) = (1 + P_o(s)C(s))^{-1}$ is the nominal sensitivity function and $W_1(s) := \gamma_r^{-1}W_r(s)$. Since the magnitude of $\Delta_P(j\omega)$ is bounded by $|W(j\omega)|$ and its phase is arbitrary for each ω, the right hand side of the above inequality can be arbitrarily close to (but it is strictly greater than) $1 - |W(j\omega)C(j\omega)S(j\omega)|$. Hence (10.5) holds for all $P \in \mathcal{P}_\Delta$ if and only if

$$|W_1(j\omega)S(j\omega)| + |W_2(j\omega)T(j\omega)| \leq 1 \qquad \text{for all } \omega \qquad (10.6)$$

where $T = 1 - S$ is the nominal complementary sensitivity function and $W_2(s) = W_m(s)$ is the multiplicative uncertainty bound $|W_m(j\omega)| > |P_o^{-1}(j\omega)\Delta_P(j\omega)|$. In the literature, $W_1(s)$ is called the performance weight and $W_2(s)$ is called the robustness weight.

In summary, a controller C achieves robust performance if (C, P_o) is stable and (10.6) is satisfied for the given weights $W_1(s)$ and $W_2(s)$. Any stabilizing controller satisfies the robust performance condition for some *specially defined* weights. For example, define $W_1(s) = \frac{\Gamma(s)}{S(s)}$ and $W_2(s) = \frac{\Upsilon(s)}{T(s)}$ with any Γ and Υ satisfying $|\Gamma(j\omega)| + |\Upsilon(j\omega)| = 1$, then (10.6) is automatically satisfied. Hence, we can always redefine W_1 and W_2 to assign more weight to performance (larger W_1) by reducing the robustness weight W_2 and vice versa. But, we should emphasize that the weights defined from an arbitrary stabilizing controller may not be very meaningful. The controller should be designed for weights given *a priori*. More precisely, the optimal control problem associated with robust performance is the following: given a nominal plant $P_o(s)$ and two weights $W_1(s)$, $W_2(s)$, find a controller $C(s)$ such that (C, P_o) is

stable and

$$|W_1(j\omega)S(j\omega)| + |W_2(j\omega)T(j\omega)| \leq \widehat{\gamma} \quad \forall \ \omega \qquad (10.7)$$

holds for the smallest possible $\widehat{\gamma}$. It is clear that the robust performance condition is automatically satisfied for the weights $\widehat{W}_1(s) = \widehat{\gamma}^{-1}W_1(s)$ and $\widehat{W}_2(s) = \widehat{\gamma}^{-1}W_2(s)$. So, by trying to minimize $\widehat{\gamma}$ we maximize the amount of allowable uncertainty magnitude and minimize the worst error energy in the tracking problem.

Example 10.2 Let $P_o(s) = \frac{(s^2+2s+2)}{s^2(s^2+4s+6)}$. Here we want to find the optimal proportional controller $C(s) = K$ such that

$$J(K) := \sup_{\omega} (|S(j\omega)| + |T(j\omega)|) \leq \widehat{\gamma}$$

for the smallest possible $\widehat{\gamma}$, subject to the condition that the feedback system (C, P_o) is stable. Applying the Routh-Hurwitz test we find that K must be positive for closed-loop system stability. Using the Bode plots, we can compute $J(K)$ for each $K > 0$. See Figure 10.6 where the smallest $\widehat{\gamma} = 2.88$ is obtained for the optimal $K = K_{\text{opt}} = 7.54$. Magnitude plots for $|S(j\omega)|$ and $|T(j\omega)|$ are shown in Figure 10.7 for $K_{\text{opt}} = 7.54$, $K_1 = 4.67$ and $K_2 = 10.41$.

In general, the problem of minimizing $\widehat{\gamma}$ over all stabilizing controllers, is more difficult than the problem of finding a controller stabilizing (C, P_o) and minimizing γ in

$$|W_1(j\omega)S(j\omega)|^2 + |W_2(j\omega)T(j\omega)|^2 \leq \gamma^2 \quad \forall \ \omega \ . \qquad (10.8)$$

A solution procedure for (10.8) is outlined in Section 10.5. In the literature, this problem is known as the mixed sensitivity minimization, or two-block \mathcal{H}^∞ control problem.

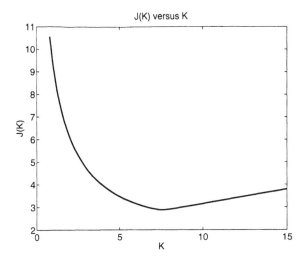

Figure 10.6: $J(K)$ versus K.

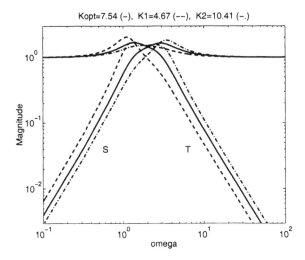

Figure 10.7: $|S(j\omega)|$ and $|T(j\omega)|$.

Exercise: Show that if $\gamma = \frac{1}{\sqrt{2}}\widehat{\gamma}$, then (10.8) implies (10.7).

Sometimes robust stability and nominal performance would be sufficient for certain applications. This is weaker (easier to satisfy) than robust performance. The nominal performance condition is

$$\|W_1 S\|_\infty \leq 1 \qquad\qquad\qquad (10.9)$$

and the robust stability condition is

$$\|W_2 T\|_\infty \leq 1. \qquad\qquad\qquad (10.10)$$

10.4 Controller Design for Stable Plants

10.4.1 Parameterization of all Stabilizing Controllers

The most important preliminary condition for optimal control problems such as (10.7) and (10.8) is stability of the nominal feedback system (C, P_o). Now we will see a description of the set of all C stabilizing the nominal feedback system (C, P_o).

Theorem 10.2 Let P_o be a _stable_ plant. Then the set of all controllers stabilizing the nominal feedback system is

$$\mathcal{C} = \left\{ C(s) = \frac{Q(s)}{1 - P_o(s)Q(s)} \quad : \quad Q(s) \text{ is proper and stable} \right\}. \ (10.11)$$

Proof: Recall that the closed-loop system is stable if and only if $S = (1 + P_o C)^{-1}$, CS and $P_o S$ are stable transfer functions. Now, if $C = Q(1 - P_o Q)^{-1}$ for some stable Q, then $S = (1 - P_o Q)$, $CS = Q$ and $P_o S = P_o(1 - P_o Q)$ are stable. Conversely, assume (C, P_o) is a stable feedback system, then in particular $C(1 + P_o C)^{-1}$ is stable. If we let

$Q = C(1 + P_oC)^{-1}$, then we get $C = Q(1 - P_oQ)^{-1}$. This completes the proof.

For unstable plants, a similar parameterization can be obtained, but in that case, derivation of the controller expression is slightly more complicated, see Section 11.4.3. A version of this controller parameterization for possibly unstable MIMO plants was obtained by Youla *et al.*, [53]. So, in the literature, the formula (10.11) is called Youla parameterization. A similar parameterization for stable nonlinear plants was obtained in [2].

10.4.2 Design Guidelines for $Q(s)$

An immediate application of the Youla parameterization is the robust performance problem. For a given stable plant P_o, with performance weight W_1 and robustness weight W_2, we want to find a controller C in the form $C = Q(1 - P_oQ)^{-1}$, where Q is stable, such that

$$|W_1(j\omega)S(j\omega)| + |W_2(j\omega)T(j\omega)| \leq 1 \quad \forall \ \omega.$$

In terms of the free parameter $Q(s)$, sensitivity and complementary sensitivity functions are:

$$S(s) = 1 - P_o(s)Q(s) \quad \text{and} \quad T(s) = 1 - S(s) = P_o(s)Q(s).$$

Hence, for robust performance, we need to design a proper stable $Q(s)$ satisfying the following inequality for all ω

$$\boxed{|W_1(j\omega)(1 - P_o(j\omega)Q(j\omega))| + |W_2(j\omega)P_o(j\omega)Q(j\omega)| \leq 1 \ .} \quad (10.12)$$

Minimum phase plants

In the light of (10.12), we design the controller by constructing a proper stable $Q(s)$ such that

- $Q(j\omega) \approx P_o^{-1}(j\omega)$ when $|W_1(j\omega)|$ is large and $|W_2(j\omega)|$ is small,

- $|Q(j\omega)| \approx 0$ when $|W_1(j\omega)| \approx 0$ and $|W_2(j\omega)|$ is large.

Obviously, if $|W_1(j\omega)|$ and $|W_2(j\omega)|$ are both large, it is impossible to achieve robust performance. In practice, W_1 is large at low-frequencies (for good tracking of low-frequency reference inputs) and W_2 is large at high frequencies for robustness against high-frequency unmodeled dynamics. This design procedure is similar to loop shaping when the plant P_o is minimum phase.

Example 10.3 Consider the following problem data:

$$P_o(s) = \frac{2}{s^2 + 2s + 4} , \quad W_1(s) = \frac{1}{s} , \quad W_2(s) = \frac{s(1 + 0.1\ s)}{10}.$$

Let $Q(s)$ be in the form

$$Q(s) = \frac{s^2 + 2s + 4}{2} \ \tilde{Q}(s)$$

where $\tilde{Q}(s)$ is a proper stable transfer function whose relative degree is at least two. The design condition (10.12) is equivalent to

$$\Phi(\omega) := |\frac{(1 - \tilde{Q}(j\omega))}{j\omega}| + |\frac{j\omega\ (1 + j\ 0.1\ \omega)\ \tilde{Q}(j\omega)}{10}| < 1 \quad \forall\ \omega .$$

So we should try to select $\tilde{Q}(j\omega) \cong 1$ for all $\omega \leq \omega_{\text{low}}$ and $|\tilde{Q}(j\omega)| \ll 1$ for all $\omega \geq \omega_{\text{high}}$, where ω_{low} can be taken as the bandwidth of the filter $W_1(s)$, i.e $\omega_{\text{low}} = \sqrt{2}$ rad/sec, and ω_{high} can be taken as the bandwidth of $W_2(s)^{-1}$, i.e. $\omega_{\text{high}} = 10$ rad/sec. Following these guidelines, let us choose $\tilde{Q}(s) = (1 + \tau\ s)^{-2}$ with τ^{-1} being close to the midpoint of the interval $(\sqrt{2}\ , 10)$, say $\tau = 1/5$. For $\tilde{Q}(s) = (1 + s/5)^{-2}$ the function $\Phi(\omega)$ is as shown in Figure 10.8: $\Phi(\omega) < 0.55 < 1$ for all ω hence, robust performance condition is satisfied. Figure 10.9 shows $J(\tau) := \max_\omega \Phi(\omega)$ as a function of $1/\tau$. We see that for all values of

$\tau^{-1} \in (2, 10)$ robust performance condition is satisfied and the best value of τ that minimizes $J(\tau)$ is $\tau = 1/5.4$, as expected, this is close to $\tau = 1/5$.

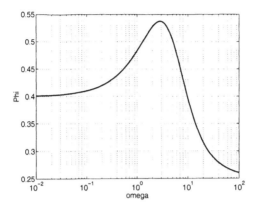

Figure 10.8: $\Phi(\omega)$ versus ω for $\tau = 1/5$.

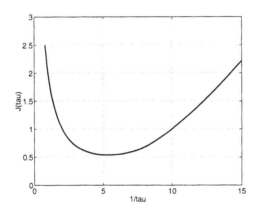

Figure 10.9: $J(\tau) := \max_\omega \Phi(\omega)$ versus $1/\tau$.

Non-minimum phase plants

When $P_o(s)$ has right half plane zeros or contains a time delay, we must be more careful. For example, let $P_o(s) = e^{-hs}\frac{(s-z)}{(s^2+2s+2)^2}$ with some $h > 0$ and $z > 0$. For good nominal performance we are tempted to choose $Q(s) \cong P_o(s)^{-1} = e^{+hs}\frac{(s^2+2s+2)^2}{(s-z)}$. There are three problems with this choice: (i) $Q(s)$ is improper, (ii) there is a time advance term in $Q(s)$ that makes it non-causal, and (iii) the pole at $z > 0$ makes $Q(s)$ unstable. The first problem can be avoided by multiplying $Q(s)$ by a strictly proper term, e.g., $\widetilde{Q}(s) = (1 + \tau s)^{-3}$ with $\tau > 0$, as demonstrated in the above example. The second and third problems can be circumvented by including the time delay and the right half plane zeros of the plant in $T(s) = P_o(s)Q(s)$.

We now illustrate this point in the context of *robust stability with asymptotic tracking*: here we want to find a stabilizing controller satisfying $\|W_2 T\|_\infty \leq 1$ and resulting in zero steady state error for unit step reference input, as well as sinusoidal reference inputs with frequency ω_o. As seen in Section 3.2, $G(s) = P_o(s)C(s)$ must contain poles at $s = 0$ and at $s = \pm j\omega_o$ for asymptotic tracking. At the poles of $G(s)$ the complementary sensitivity $T(s) = G(s)(1+G(s))^{-1}$ is unity. Therefore, we must have

$$T(s)\big|_{s=0,\pm j\omega_o} = 1 \ . \tag{10.13}$$

But the controller parameterization implies $T(s) = P_o(s)Q(s)$. So, robust stability condition is satisfied only if $|W_2(0)| \leq 1$ and $|W_2(\pm j\omega_o)| \leq 1$. Furthermore, we need $P_o(0) \neq 0$ and $P_o(\pm j\omega_o) \neq 0$, otherwise (10.13) does not hold for any stable $Q(s)$. If $P_o(s)$ contains a time delay, e^{-hs}, and zeros z_1, \ldots, z_k in the right half plane, then $T(s)$ must contain at least the same amount of time delay and zeros at z_1, \ldots, z_k, because $Q(s)$ cannot have any time advance or poles in the right half

plane. To be more precise, let us factor $P_o(s)$ as

$$P_o(s) = P_{ap}(s)P_{mp}(s) , \quad P_{ap}(s) = e^{-hs} \prod_{i=1}^{k} \frac{z_i - s}{z_i + s} \quad (10.14)$$

and $P_{mp}(s) = P_o(s)/P_{ap}(s)$ is minimum phase. Then, $T(s) = P_o(s)Q(s)$ must be in the form

$$T(s) = P_{ap}(s)\widetilde{T}(s)$$

where $\widetilde{T} = P_{mp}Q$ is a stable transfer function whose relative degree is at least the same as the relative degree of $P_{mp}(s)$,

$$\widetilde{T}(s)\big|_{s=0,\pm j\omega_o} = \frac{1}{P_{ap}(s)}\big|_{s=0,\pm j\omega_o}$$

and $\|W_2\widetilde{T}\|_\infty \leq 1$. Once $\widetilde{T}(s)$ is determined, the controller parameter $Q(s)$ can be chosen as $Q(s) = \widetilde{T}(s)/P_{mp}(s)$ (note that Q is proper and stable) and then the controller becomes

$$C(s) = \frac{Q(s)}{1 - P_o(s)Q(s)} = \frac{\widetilde{T}(s)/P_{mp}(s)}{1 - P_{ap}(s)\widetilde{T}(s)}. \quad (10.15)$$

The controller itself may be unstable, but the feedback system is stable with the choice (10.15).

Example 10.4 Here, we consider

$$P_o(s) = \frac{(s - 3)}{(s^2 + 2s + 2)}\cdot\cdot$$

So, $P_{ap}(s) = \frac{3-s}{3+s}$ and $P_{mp}(s) = \frac{-(s+3)}{(s^2+2s+2)}$. We want to design a controller such that the feedback system is robustly stable for $W_2(s) = 0.5(1 + s/2)$ and steady state tracking error is zero for step reference inputs as well as sinusoidal reference inputs whose periods are 2π, i.e., $\omega_o = 1$ rd/sec.

There are four design conditions for $\widetilde{T}(s)$:

(i) $\widetilde{T}(s)$ must be stable,

(ii) $\widetilde{T}(0) = (P_{\mathrm{ap}}(0))^{-1} = 1$, $\widetilde{T}(\pm j1) = (P_{\mathrm{ap}}(\pm j1))^{-1} = 0.8 \pm j0.6$,

(iii) relative degree of $\widetilde{T}(s)$ is at least one,

(iv) $|\widetilde{T}(j\omega)| \leq 2(1 + \omega^2/4)^{-\frac{1}{2}}$ for all ω.

We may assume that

$$\widetilde{T}(s) = \frac{a_2 s^2 + a_1 s + a_0}{(s+x)^3} + \frac{s(s^2+1)}{(s+x)^3} \widetilde{T}_0(s)$$

for some $x > 0$; the coefficients a_2, a_1, a_0 are to be determined from the interpolation conditions imposed by (ii), and $\widetilde{T}_0(s)$ is an arbitrary stable transfer function. Interpolation conditions (ii) are satisfied if

$$a_0 = x^3$$

$$a_1 = \mathrm{Im}(\,(0.8 + j0.6)(x + j1)^3\,)$$

$$a_2 = x^3 - \mathrm{Re}(\,(0.8 + j0.6)(x + j1)^3\,) \,.$$

Let us try to find a feasible $\widetilde{T}(s)$ by arbitrarily picking $\widetilde{T}_0(s) = 0$ and testing the fourth design condition for different values of $x > 0$. By using MATLAB, we can compute $\|W_2\widetilde{T}\|_\infty$ for each fixed $x > 0$. See Figure 10.10, which shows that the robustness condition (iv) is satisfied for $x \in (0.456\ ,\ 1.027)$. The best value of x (in the sense that $\|W_2 T\|_\infty$ is minimum) is $x = 0.62$.

Exercises:

1. For $x = 0.62$, compute $\widetilde{T}(s)$ and the associated controller (10.15), verify that the controller has poles at $s = 0, \pm j1$. Plot $|W_2(j\omega)T(j\omega)|$ versus ω and check that the robustness condition is satisfied.

2. Find the best value of $x \in (0.456\ ,\ 1.027)$ which maximizes the vector margin.

3. Define $\widetilde{T}_0(s) := -\frac{(a_2 s^2 + a_1 s + a_0)}{(s+\epsilon)(s^2 + \epsilon s + 1)}$, and pick $x = 0.62$. Plot $|T(j\omega)|$ versus ω for $\epsilon = 0.2,\ 0.1,\ 0.01,\ 0.001$. Find $\|W_2 T\|_\infty$ for the same

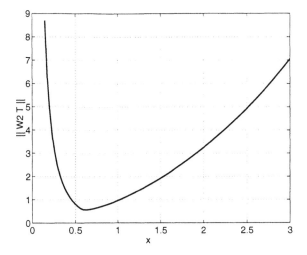

Figure 10.10: $\|W_2\widetilde{T}\|_\infty$ versus x.

values of ϵ and compare these results with the lower bound

$$\max\{|W_2(0)| \, , \, |W_2(\pm j1)|\} \le \|W_2T\|_\infty.$$

Smith predictor

For stable time delay systems there is a popular controller design method called Smith predictor, which dates back to 1950s. We will see a connection between the controller parameterization (10.11) and the Smith predictor dealing with stable plants in the form $P_o(s) = P_0(s)e^{-hs}$ where $h > 0$ and $P_0(s)$ is the nominal plant model for $h = 0$. To obtain a stabilizing controller for $P_o(s)$, first design a controller $C_0(s)$ that stabilizes the non-delayed feedback system (C_0, P_0). Then let $Q_0 := C_0(1 + P_0C_0)^{-1}$; note that by construction $Q_0(s)$ is stable. If we use $Q = Q_0$ in the parameterization of stabilizing controllers (10.11) for $P_o(s) = P_0(s)e^{-hs}$, we obtain the following controller

$$C(s) = \frac{Q_0(s)}{1 - e^{-hs}P_0(s)Q_0(s)} = \frac{C_0(s)}{1 + P_0(s)C_0(s)(1 - e^{-hs})}. \qquad (10.16)$$

The controller structure (10.16) is called Smith predictor, [48]. When C is designed according to (10.16), the complementary sensitivity function is $T(s) = e^{-hs}T_0(s)$ where $T_0(s)$ is the complementary sensitivity function for the non-delayed system: $T_0(s) = \frac{P_0(s)C_0(s)}{1+P_0(s)C_0(s)}$. Therefore, $|T(j\omega)| = |T_0(j\omega)|$ for all ω. This fact can be exploited to design robustly stabilizing controllers. For a given multiplicative uncertainty bound W_2, determine a controller C_0 from P_0 satisfying $\|W_2T_0\|_\infty \le 1$. Then, the controller (10.16) robustly stabilizes the feedback system with nominal plant $P_o(s) = P_0(s)e^{-hs}$ and multiplicative uncertainty bound $W_2(s)$. The advantage of this approach is that C_0 is designed independent of time delay; it only depends on W_2 and P_0. Also, note that the poles of $T(s)$ are exactly the poles of the non-delayed system $T_0(s)$.

10.5 Design of \mathcal{H}^∞ Controllers

A solution for the \mathcal{H}^∞ optimal mixed sensitivity minimization problem (10.8) is presented in this section. For MIMO finite dimensional systems a state space based solution is available in MATLAB: the relevant command is hinfsyn (it assumes that the problem data is transformed to a "generalized" state space form that includes the plant and the weights), see [4]. The solution procedure outlined here is taken from [50]. It allows infinite dimensional plant models and uses SISO transfer function representations.

10.5.1 Problem Statement

Let us begin by restating the \mathcal{H}^∞ control problem we are dealing with: given a nominal plant model $P_o(s)$ and two weighting functions $W_1(s)$ and $W_2(s)$, we want to find a controller C stabilizing the feedback system (C, P_o) and minimizing $\gamma > 0$ in

$$\Psi(\omega) := |W_1(j\omega)S(j\omega)|^2 + |W_2(j\omega)T(j\omega)|^2 \le \gamma^2 \qquad \forall\ \omega$$

where $S = (1 + P_o C)^{-1}$ and $T = 1 - S$. The optimal controller is denoted by C_{opt} and the resulting minimal γ is called γ_{opt}.

Once we find an arbitrary controller that stabilizes the feedback system, then the peak value of the corresponding $\Psi(\omega)$ is an upper bound for γ_{opt}^2. It turns out that for the optimal controller we have $\Psi(\omega) = \gamma_{\text{opt}}^2$ for all ω. This fact implies that if we find a stabilizing controller for which $\Psi(\omega)$ is not constant, then the controller is not optimal, i.e., the peak value of $\Psi(\omega)$ can be reduced by another stabilizing controller.

Assumptions and preliminary definitions

Let $\{z_1, \ldots, z_k\}$ be the zeros and $\{p_1, \ldots, p_\ell\}$ be the poles of $P_o(s)$ in the open right half plane. Suppose that $P_o(s) = e^{-hs} P_0(s)$, where $P_0(s)$ is rational and $h \geq 0$. Define the following proper stable functions

$$M_n(s) := e^{-hs} \prod_{i=1}^{k} \frac{z_i - s}{z_i + s}, \quad M_d(s) := \prod_{i=1}^{\ell} \frac{p_i - s}{p_i + s}, \quad N_o(s) := \frac{P_o(s) M_d(s)}{M_n(s)}.$$

Note that $M_n(s)$ and $M_d(s)$ are all-pass and $N_o(s)$ is minimum phase. Since $P_0(s)$ is a rational function, $N_o(s)$ is in the form

$$N_o(s) = \frac{nN_o(s)}{dN_o(s)}$$

for some stable polynomials $nN_o(s)$ and $dN_o(s)$. Moreover, $|P_o(j\omega)| = |N_o(j\omega)|$ for all ω. As usual, we assume $P_o(s)$ to be strictly proper, which means that $\deg(nN_o) < \deg(dN_o)$.

For example, when

$$P_o(s) = e^{-3s} \frac{5(s-1)^2(s+3)}{(s-2)(s^2 + 2s + 2)^2},$$

we define

$$M_n(s) = e^{-3s} \frac{(1-s)^2}{(1+s)^2}, \quad M_d(s) = \frac{(2-s)}{(2+s)},$$

$$N_o(s) = \frac{-5(s+1)^2(s+3)}{(s+2)(s^2+2s+2)^2}.$$

We assume that $W_1(s)$ is in the form $W_1(s) = nW_1(s)/dW_1(s)$, where $nW_1(s)$ and $dW_1(s)$ are stable polynomials with $\deg(nW_1) = \deg(dW_1) = n_1 \geq 1$. We will also use the notations

$$M_1(s) := \frac{dW_1(-s)}{dW_1(s)},$$

$$E(s) := \frac{nW_1(-s)}{dW_1(s)} - \gamma^2 \frac{dW_1(-s)}{nW_1(s)}.$$

Recall that since $W_2(s) = W_m(s)$ we have $|W_2(j\omega)| = \frac{|W(j\omega)|}{|P_o(j\omega)|}$, where $|W(j\omega)|$ is the upper bound of the additive plant uncertainty magnitude. So we let $W_2(s) = W(s)N_o(s)^{-1}$ with $W(s) = nW(s)/dW(s)$, where $nW(s)$ and $dW(s)$ are stable polynomials and we assume that $\deg(nW) = \deg(dW) \geq 0$.

10.5.2 Spectral Factorization

Let $A(s)$ be a transfer function such that $A(j\omega)$ is real and $A(j\omega) \geq \delta$ for all ω, for some $\delta > 0$. Then, there exists a proper stable function $B(s)$ such that $B(s)^{-1}$ is also proper and stable and $|B(j\omega)|^2 = A(j\omega)$. Construction of $B(s)$ from $A(j\omega)$ is called spectral factorization. There are several optimal control problems whose solutions require spectral factorizations; \mathcal{H}^∞ control is one of them.

For the solution of mixed sensitivity minimization problem define

$$A(j\omega) := \left(|W(j\omega)|^2 + |W_1(j\omega)|^2 \left(|N_o(j\omega)|^2 - \frac{|W(j\omega)|^2}{\gamma^2} \right) \right)^{-1}.$$

Since $W_1(s)$, $W(s)$ and $N_o(s)$ are rational functions, we can write $A(s) = nA(s)/dA(s)$ where $nA(s)$ and $dA(s)$ are polynomials:

$$nA(s) = dW_1(s)dW_1(-s)dN_o(s)dN_o(-s)dW(s)dW(-s)$$

$$
\begin{aligned}
dA(s) \;=\; & nW_1(-s)nW_1(s)nN_o(s)nN_o(-s)dW(s)dW(-s) \\
& + dW_1(-s)dW_1(s)dN_o(s)dN_o(-s)nW(s)nW(-s) \\
& - \gamma^{-2}nW_1(-s)nW_1(s)dN_o(s)dN_o(-s)nW(s)nW(-s).
\end{aligned}
$$

By symmetry, if $\hat{p} \in \mathbb{C}$ is a root of $dA(s)$, then so is $-\hat{p}$. Suppose γ is such that $dA(s)$ has no roots on the Im-axis. Then we can label the roots of $dA(s)$ as $\hat{p}_1, \ldots, \hat{p}_{2n_b}$ with $\hat{p}_{n_b+i} = -\hat{p}_i$, and $\mathrm{Re}(\hat{p}_i) < 0$ for all $i = 1, \ldots, n_b$. Finally, the spectral factor $B(s) = nB(s)/dB(s)$ can be determined as

$$
\begin{aligned}
nB(s) \;&=\; dW_1(s)\, dN_o(s)\, dW(s) \\
dB(s) \;&=\; \sqrt{dA(0)} \prod_{i=1}^{n_b}(1 - s/\hat{p}_i).
\end{aligned}
$$

Note that $B(s)$ is unique up to multiplication by -1 (i.e. $B(s)$ and $-B(s)$ have the same poles and zeros, and the same magnitude on the Im-axis). Another important point to note is that both $B(s)$ and $B(s)^{-1}$ are proper stable functions.

10.5.3 Optimal \mathcal{H}^∞ Controller

The optimal \mathcal{H}^∞ controller can be expressed in the form

$$
C_{\mathrm{opt}}(s) = \frac{W_1(s)}{\gamma_{\mathrm{opt}}} \frac{B(s)M_d(s)E(s)}{L_{\mathrm{opt}}(s) + \gamma_{\mathrm{opt}} M_1(s)M_n(s)N_o(s)B(s)} \tag{10.17}
$$

where $L_{\mathrm{opt}}(s) = nL(s)/dL(s)$, for some polynomials $nL(s)$ and $dL(s)$ with $\deg(nL) = \deg(dL) \le (n_1 + \ell - 1)$. The function $L_{\mathrm{opt}}(s)$ is such that $D_{\mathrm{opt}}(s)$ and $\widehat{D}_{\mathrm{opt}}(s)$ do not have any poles at the closed right half plane zeros of $M_d(s)$ and $E(s)$:

$$
D_{\mathrm{opt}}(s) \;:=\; \frac{L_{\mathrm{opt}}(s) + \gamma_{\mathrm{opt}} M_1(s)M_n(s)N_o(s)B(s)}{B(s)M_d(s)E(s)} \tag{10.18}
$$

$$
\widehat{D}_{\mathrm{opt}}(s) \;:=\; \frac{L_{\mathrm{opt}}(s) - 1/L_{\mathrm{opt}}(-s)}{B(s)M_d(s)E(s)} \tag{10.19}
$$

Next, we compute $L_{\mathrm{opt}}(s)$ and γ_{opt} from these interpolation conditions.

First, note that the zeros of $E(s)$ can be labeled as $\hat{z}_1, \ldots, \hat{z}_{2n_1}$, with $\hat{z}_{n_1+i} = -\hat{z}_i$ and $\text{Re}(\hat{z}_i) \geq 0$ for all $i = 1, \ldots, n_1$. Now define

$$\alpha_i := \begin{cases} p_i & i = 1, \ldots, \ell \\ \hat{z}_{i-\ell} & i = \ell+1, \ldots, \ell+n_1 =: n \end{cases}$$

where p_i, $i = 1, \ldots, \ell$ are the zeros of $M_d(s)$ (i.e., unstable poles of the plant). For simplicity, assume that $\alpha_i \neq \alpha_j$ for $i \neq j$. Total number of unknown coefficients to be determined for $nL(s)$ and $dL(s)$ is $2n$. Because of the symmetric interpolation points, it turns out that $nL(s) = dL(-s)$ and hence $\hat{D}_{\text{opt}}(s) \equiv 0$. Let us introduce the notation

$$dL(s) = \sum_{i=0}^{n-1} v_i s^i = [1 \quad s \quad \cdots \quad s^{n-1}] \, v \qquad (10.20)$$

$$nL(s) = \sum_{i=0}^{n-1} v_i(-s)^i = [1 \quad s \quad \cdots \quad s^{n-1}] \, J_n \, v \qquad (10.21)$$

where $v := [v_0 \ \ldots v_{n-1}]^\mathsf{T}$ denotes the vector of unknown coefficients of $dL(s)$ and J_n is an $n \times n$ diagonal matrix whose ith diagonal entry is $(-1)^{i+1}$, $i = 1, \ldots, n$. For any $m \geq 1$ define the $n \times m$ Vandermonde matrix

$$V_m := \begin{bmatrix} 1 & \alpha_1 & \cdots & \alpha_1^{m-1} \\ \vdots & \vdots & \cdots & \vdots \\ 1 & \alpha_n & \cdots & \alpha_n^{m-1} \end{bmatrix},$$

and the $n \times n$ diagonal matrix F whose ith diagonal entry is $F(\alpha_i)$ for $i = 1, \ldots, n$ where

$$F(s) := \gamma M_1(s) M_n(s) N_o(s) B(s).$$

Finally, γ_{opt} is the largest value of γ for which the $n \times n$ matrix

$$R := V_n J_n + F \, V_n$$

is singular. By plotting the smallest singular value of R (in MATLAB's notation $\min(\text{svd}(R))$) versus γ we can detect γ_{opt}. The corresponding singular vector v_{opt} satisfies

$$R_{\text{opt}} v_{\text{opt}} = 0 \tag{10.22}$$

where R_{opt} is the matrix R evaluated at $\gamma = \gamma_{\text{opt}}$. The vector v_{opt} defines $L_{\text{opt}}(s)$ via (10.20) and (10.21). The optimal controller $C_{\text{opt}}(s)$ is determined from $L_{\text{opt}}(s)$ and γ_{opt}.

The controller (10.17) can be rewritten as

$$C_{\text{opt}}(s) = \frac{W_1(s)}{d_{\text{opt}}\gamma_{\text{opt}}} \frac{1}{1 + H_{\text{opt}}(s)} \tag{10.23}$$

$$H_{\text{opt}}(s) := \frac{D_{\text{opt}}(s)}{d_{\text{opt}}} - 1 \tag{10.24}$$

where $D_{\text{opt}}(s)$ is given by (10.18), and

$$d_{\text{opt}} := D_{\text{opt}}(\infty) = \frac{W_1(\infty) \, |W(\infty)|/\gamma_{\text{opt}}^2}{\sqrt{1 - |W_1(\infty)|^2/\gamma_{\text{opt}}^2}}.$$

Clearly $H_{\text{opt}}(s)$ is strictly proper, moreover it is stable if $L_{\text{opt}}(s)$ is stable. Controller implementation in the form (10.23) is preferred over (10.17) because when $M_n(s)$ contains a time delay term, it is easier to approximate $H_{\text{opt}}(s)$ than $D_{\text{opt}}(s)$.

Example 10.5 We now apply the above procedure to the following problem data:

$$P_o(s) = \frac{e^{-hs}}{(s-1)}, \quad W_1(s) = \frac{(10+s)}{(1+10s)}, \quad W(s) = 0.1 .$$

Our aim is to find γ_{opt} and the corresponding \mathcal{H}^∞ optimal controller $C_{\text{opt}}(s)$. We begin by defining

$$M_n(s) = e^{-hs}, \quad M_d(s) = \frac{(1-s)}{(1+s)}, \quad N_o(s) = \frac{-1}{(1+s)}$$

$$M_1(s) = \frac{(1-10s)}{(1+10s)}, \quad E(s) = \frac{(100-\gamma^2)+(100\gamma^2-1)s^2}{(1+10s)(10+s)}.$$

Next, we perform the spectral factorization for

$$dA(s) = (100.01 - \gamma^{-2}) - (2.01 - 1.01\gamma^{-2})s^2 + (1 - 0.01\gamma^{-2})s^4.$$

Since $dB(s)dB(-s) = dA(s)$ we define $dB(s) = (a + bs + cs^2)$, where
$a = \sqrt{100.01 - \gamma^{-2}}$, $c = \sqrt{1 - 0.01\gamma^{-2}}$, $b = \sqrt{2.01 - 1.01\gamma^{-2} + 2ac}$
that leads to

$$B(s) = \frac{(1 + 10s)(1 + s)}{(a + bs + cs^2)} \quad \text{and} \quad F(s) = \frac{\gamma e^{-hs}(10s - 1)}{(a + bs + cs^2)}.$$

Let γ_{\min} be the smallest and γ_{\max} be the largest γ values for which this spectral factorization makes sense, i.e., $a > 0$, $b > 0$, $c > 0$. Then, γ_{opt} lies in the interval $(\gamma_{\min}, \gamma_{\max})$. A simple algebra shows that $\gamma_{\min} = 0.2341$, and $\gamma_{\max} = \sqrt{100.01} \cong 10$. For $\gamma \in (0.2341, 10)$ the zeros of $E(s)$ are $\hat{z}_{1,2} = \pm j\sqrt{\frac{100-\gamma^2}{100\gamma^2-1}}$ and the only zero of $M_d(s)$ is $p_1 = 1$. Therefore, we may choose $\alpha_1 = 1$ and $\alpha_2 = j\sqrt{\frac{100-\gamma^2}{100\gamma^2-1}}$. Then, the 2×2 matrix R is

$$R = \begin{bmatrix} 1 & \alpha_1 \\ 1 & \alpha_2 \end{bmatrix} \begin{bmatrix} 1 & 0 \\ 0 & -1 \end{bmatrix} - \begin{bmatrix} F(\alpha_1) & 0 \\ 0 & F(\alpha_2) \end{bmatrix} \begin{bmatrix} 1 & \alpha_1 \\ 1 & \alpha_2 \end{bmatrix}.$$

The smallest singular value of R versus γ plots are shown in Figure 10.11 for three different values of $h = 0.1, 0.5, 0.95$. We see that the optimal γ values are $\gamma_{\text{opt}} = 0.56, 1.23, 2.49$, respectively; (since the zeros of $E(s)$ are not distinct at $\gamma = 10$, the matrix R becomes singular, independent of h at this value of γ; this is the reason that we discard $\gamma = 10$). The optimal controller for $h = 0.1$ is determined from $\gamma_{\text{opt}} = 0.56$ and the corresponding $v_{\text{opt}} = [-0.4739 \quad -0.8806]^T$, which gives $dL(s) = -(0.4739 + 0.8806\ s)$. Since $nL(s) = dL(-s)$ we have $L_{\text{opt}}(s) \cong (1 - 1.86s)/(1 + 1.86s)$. Note that $L_{\text{opt}}(s)$ is stable, so the resulting $H_{\text{opt}}(s)$ is stable as well. The Nyquist plot of $H_{\text{opt}}(j\omega)$ is shown in Figure 10.12, from which we deduce that $C_{\text{opt}}(s)$ is stable.

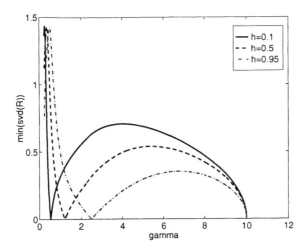

Figure 10.11: $\mathtt{min}(\mathtt{svd}(\mathrm{R}))$ versus γ.

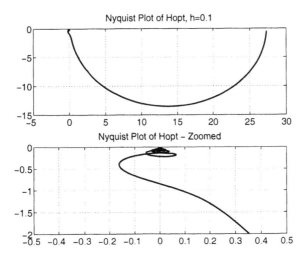

Figure 10.12: Nyquist plot of $H_{\mathrm{opt}}(j\omega)$ for $h = 0.1$.

Example 10.6 Here we consider a finite dimensional plant model

$$P_o(s) = \frac{(2-s)}{(2+s)(1+s)}$$

with the weights $W_1(s)$ and $W(s)$ being the same as in the previous example. In this case, the plant is stable and $W_1(s)$ is first order, so $n = n_1 + \ell = 1$, which implies that $L_{opt}(s) = 1$. The matrix R is 1×1 and γ_{opt} is the largest root of

$$R = 1 + F(\alpha) = 0 \qquad\qquad (10.25)$$

where α is equal to α_2 of the previous example and

$$F(s) = \gamma \frac{(1 - 10s)(2-s)}{(2+s)(a + bs + cs^2)}$$

a, b, c being the same as above. The root of equation (10.25) in $\gamma \in$ $(0.2341\ ,\ 10)$ is $\gamma_{opt} \cong 0.83$. After internal pole zero cancelations within the optimal controller, we get

$$C_{opt}(s) \cong 8.3 \,\frac{(s+2)(s+1)}{(s+15)(s+0.1)}.$$

10.5.4 Suboptimal \mathcal{H}^∞ Controllers

In this section, we characterize the set of all suboptimal \mathcal{H}^∞ controllers stabilizing the feedback system (C, P_o) and achieving $\Psi(\omega) \leq \gamma^2$ for a given suboptimal performance level $\gamma > \gamma_{opt}$. All suboptimal controllers have the same structure as the optimal controller:

$$
\begin{aligned}
C_{sub}(s) &= \frac{W_1(s)}{\gamma\, d_{sub}} \frac{1}{1 + H_{sub}(s)} \\[2mm]
H_{sub}(s) &:= \frac{D_{sub}(s)}{d_{sub}} - 1 \\[2mm]
D_{sub}(s) &:= \frac{L_{sub}(s) + \gamma M_1(s) M_n(s) N_o(s) B(s)}{B(s) M_d(s) E(s)} \\[2mm]
d_{sub} &:= D_{sub}(\infty)\,.
\end{aligned}
$$

In this case $L_{sub}(s)$ is in the form

$$L_{sub}(s) = L_0(s)\,\frac{1 + Q(s)/\widehat{L}(-s)}{1 + \widehat{L}(s)Q(s)} \tag{10.26}$$

where $Q(s)$ is an arbitrary proper stable transfer function with $\|Q\|_\infty \leq 1$ and $L_0(s) = nL_0(s)/dL_0(s)$, $\widehat{L}(s) = nL_0(-s)/dL_0(s)$, with $\deg(nL_0) = \deg(dL_0) \leq n$. By using (10.26) in $D_{sub}(s)$ we get

$$\begin{aligned}
D_{sub}(s) &= D_0(s) - \widehat{D}(s)\,\frac{\widehat{L}(s)Q(s)}{1 + \widehat{L}(s)Q(s)} \\
D_0(s) &:= \frac{L_0(s) + \gamma M_1(s)M_n(s)N_o(s)B(s)}{B(s)M_d(s)E(s)} \\
\widehat{D}(s) &:= \frac{L_0(s) - 1/L_0(-s)}{B(s)M_d(s)E(s)}\,.
\end{aligned}$$

In particular, for $Q(s) = 0$ we have $D_{sub}(s) = D_0(s)$.

Note that, in this case there are $2(n+1)$ unknown coefficients in nL_0 and dL_0. The interpolation conditions are similar to the optimal case: $D_0(s)$ and $\widehat{D}(s)$ should have no poles at the closed right half plane zeros of $M_d(s)$ and $E(s)$. These interpolation conditions give $2(n_1 + \ell) = 2n$ equations. We need two more equations to determine nL_0 and dL_0. Assume that $nW_1(s)$ has a zero \tilde{z} that is distinct from $-\alpha_i$ for all $i = 1, \ldots, n$. Then, we can set $dL_0(\tilde{z}) = 0$ and $nL_0(\tilde{z}) = 1$. The last two equations violate the symmetry of interpolation conditions, so $nL_0(s) \neq dL_0(-s)$ and hence $\widehat{D}(s) \neq 0$. Let us define

$$\begin{aligned}
dL_0(s) &= [1 \quad s \quad \cdots \quad s^n]\,\mathsf{v} \tag{10.27} \\
nL_0(s) &= [1 \quad s \quad \cdots \quad s^n]\,\mathsf{w} \tag{10.28} \\
\tilde{z}_{n+1} &= [1 \quad \tilde{z} \quad \cdots \tilde{z}^n]
\end{aligned}$$

where $\mathsf{v} = [\mathsf{v}_0 \quad \cdots \quad \mathsf{v}_n]^\mathsf{T}$ and $\mathsf{w} = [\mathsf{w}_0 \quad \cdots \quad \mathsf{w}_n]^\mathsf{T}$ are the unknown coefficients of $dL_0(s)$ and $nL_0(s)$, respectively. The set of $2(n+1)$ equations corresponding to the above mentioned interpolation conditions can be

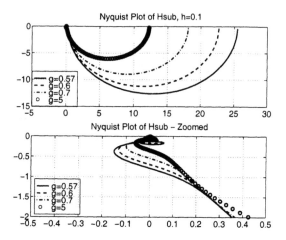

Figure 10.13: Nyquist plot of $H_{\text{sub}}(j\omega)$ for $h = 0.1$.

written as

$$
\begin{bmatrix} 0_{n\times 1} \\ 0_{n\times 1} \\ 0 \\ 1 \end{bmatrix} = \begin{bmatrix} V_{n+1} & F\,V_{n+1} \\ F\,V_{n+1}J_{n+1} & V_{n+1}J_{n+1} \\ 0_{1\times(n+1)} & \tilde{z}_{n+1} \\ \tilde{z}_{n+1} & 0_{1\times(n+1)} \end{bmatrix} \begin{bmatrix} \mathbf{w} \\ \mathbf{v} \end{bmatrix}. \tag{10.29}
$$

For $\gamma > \gamma_{\text{opt}}$ equations (10.29) yield unique vectors \mathbf{w} and \mathbf{v}. Then, by (10.27) and (10.28) we define $nL_0(s)$ and $dL_0(s)$, and thus parameterize all suboptimal \mathcal{H}^∞ controllers via $L_{\text{sub}}(s)$, (10.26).

Example 10.7 For the \mathcal{H}^∞ optimal control problem studied in Example 10.5 we have determined $C_{\text{opt}}(s)$ for $h = 0.1$. Now we find suboptimal controllers for the same plant with a performance level $\gamma > 0.56 = \gamma_{\text{opt}}$. In this case, we construct the set of equations (10.29) and solve for \mathbf{v} and \mathbf{w}, these vectors along with an arbitrarily selected proper stable $Q(s)$, with $\|Q\|_\infty \leq 1$, give $L_{\text{sub}}(s)$, which defines $C_{\text{sub}}(s)$. For $Q = 0$, i.e. $L_{\text{sub}} = L_0$, the Nyquist plots of $H_{\text{sub}}(s)$ are shown in Figure 10.13 for $\gamma = 0.57,\ 0.6,\ 0.7,\ 5$.

10.6 Exercise Problems

1. Consider the set of plants

$$\mathcal{P}_\Delta \quad := \quad \{ \, P(s) = P_o(s) + \Delta(s) \quad :$$

$$\Delta(s) \text{ is stable and } |\Delta(j\omega)| < \sqrt{2} \ \forall \, \omega \}$$

where the nominal plant transfer function is $P_o(s) = \frac{1}{s+4}$.
We want to design a controller in the form $C(s) = \frac{K}{s}$ such that

 (i) the nominal feedback system is stable with closed-loop system poles r_1 and r_2 satisfying $\mathrm{Re}(r_i) < -1$ for $i = 1, 2$; and

 (ii) the feedback system is robustly stable.

 Show that no controller exists that can satisfy both (i) and (ii).

2. A second-order plant model is in the form

$$P(s) = \frac{0.2 \, \omega_o^2}{s^2 + 2\zeta\omega_o s + \omega_o}.$$

Let $\omega_o \in (9 \, , \, 15)$ and $\zeta \in (0.1 \, , \, 0.3)$ with nominal values $\omega_o = 10$ and $\zeta = 0.15$, i.e. $P_o(s) := \frac{20}{s^2+3s+100}$. Find an appropriate $W(s)$ so that

$$|P(j\omega) - P_o(j\omega)| < |W(j\omega)| \quad \forall \, \omega \, .$$

Define $W_2(s) = W(s)P_o(s)^{-1}$ and $W_1(s) = \frac{40+s}{1+100s}$. By using the Youla parameterization find a controller achieving robust performance: $|W_1(j\omega)S(j\omega)| + |W_2(j\omega)T(j\omega)| \leq 1 \ \forall \, \omega.$

3. Let

$$P_o(s) = e^{-2s} P_0(s) \, , \qquad P_0(s) = \frac{(s - 1)}{(s + 2)(s^2 + 2s + 2)}$$

and $W_2(s) = 0.8(1+s/4)^2$. Design a robustly stabilizing controller achieving asymptotic tracking of step reference inputs. Write this controller in the Smith predictor form

$$C(s) = \frac{C_0(s)}{1 + P_0(s)C_0(s)(1 - e^{-2s})}.$$

and determine $C_0(s)$ in this expression.

4. Consider the \mathcal{H}^∞ optimal control problem data

$$P_o(s) = \frac{(z-s)}{(z+s)(s+1)}, \quad W_1(s) = \frac{(10+s)}{(1+10s)} \quad W(s) = 0.1$$

and plot γ_{opt} versus $z > 0$. Compute \mathcal{H}^∞ optimal controllers for $z = 0.2, 1, 20$. What are the corresponding closed-loop system poles?

5. Consider Example 10.5 of Section 10.5.3. Find the \mathcal{H}^∞ optimal controllers for $h = 0.5$ sec and $h = 0.95$ sec. Are these controllers stable?

6. (i) Find $L_0(s)$ and $\widehat{L}(s)$ of the suboptimal controllers determined in Example 10.7 of Section 10.5.4 for $h = 0.1$ sec with $\gamma = 0.57, 0.6, 0.7, 5$.

 (ii) Select $Q(s) = 0$ and obtain the Nyquist plot of $H_{\text{sub}}(s)$ for $h = 0.5$ sec and $\gamma = 1.25, 1.5, 7.5$. Plot $\Phi(\omega)$ versus ω corresponding to these suboptimal controllers.

 (iii) Repeat (ii) for $Q(s) = \frac{e^{-s}(s-3)}{(s+3)}$.

Chapter 11

Basic State Space
Methods

The purpose of this chapter is to introduce basic linear controller design methods involving state space representation of a SISO plant. The coverage here is intended to be an overview at the introductory level. Interested readers are referred to new comprehensive texts on linear system theory such as [3], [6], [11], [47], [54].

11.1 State Space Representations

Consider the standard feedback system shown in Figure 11.1, where the plant is given in terms of a state space realization

$$\text{Plant}: \begin{cases} \dot{x}(t) = Ax(t) + Bu(t) \\ y_o(t) = Cx(t) + Du(t) \end{cases} \qquad (11.1)$$

Here A, B, C and D are appropriate size constant matrices with real entries and $x(t) \in \mathbb{R}^n$ is the state vector associated with the plant.

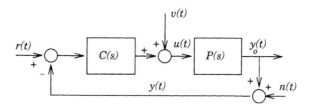

Figure 11.1: Standard feedback system.

We make the following assumptions: (i) $D = 0$, which means that the plant transfer function $P(s) = C(sI - A)^{-1}B + D$, is strictly proper, and (ii) $u(t)$ and $y_o(t)$ are scalars, i.e., the plant is SISO. The multiple output case is considered in the next section only for the specific output $y_o(t) = x(t)$.

The realization (11.1) is said to be *controllable* if the $n \times n$ controllability matrix U is invertible:

$$U := \left[B \vdots AB \vdots \cdots \vdots A^{n-1}B \right] \tag{11.2}$$

(U is obtained by stacking n vectors $A^k B$, $k = 0, \ldots, n-1$, side by side). Controllability of (11.1) depends on A and B only; so, when U is invertible, we say that the pair (A, B) is controllable. The realization (11.1) is said to be *observable* if the pair $(A^{\mathsf{T}}, C^{\mathsf{T}})$ is controllable (the superscript $^{\mathsf{T}}$ denotes the transpose of a matrix).

Unless otherwise stated, the realization (11.1) will be assumed to be controllable and observable, i.e., this is a *minimal* realization of the plant (which means that n, the dimension of the state vector, is the smallest among all possible state space realizations of $P(s)$). In this case, the poles of the plant are the roots of the polynomial $\det(sI - A)$. From any minimal realization $\{A, B, C, D\}$, another minimal realization $\{A_z, B_z, C_z, D\}$ can be obtained by defining $z(t) = Zx(t)$ as the new state vector, where Z is an arbitrary $n \times n$ invertible matrix. Note that $A_z = ZAZ^{-1}$, $B_z = ZB$ and $C_z = CZ^{-1}$. Since these two realizations

represent the same plant, they are said to be *equivalent*. In Section 2.1, we saw the controllable canonical state space realization

$$A_c := \begin{bmatrix} 0_{(n-1)\times 1} & I_{(n-1)\times(n-1)} \\ -a_n & \cdots & -a_1 \end{bmatrix} \qquad B_c := \begin{bmatrix} 0_{(n-1)\times 1} \\ 1 \end{bmatrix} \qquad (11.3)$$

$$C_c := \begin{bmatrix} b_n & \cdots & b_1 \end{bmatrix} \qquad\qquad D := d, \qquad (11.4)$$

which was derived from the transfer function

$$P(s) = \frac{N_P(s)}{D_P(s)} = \frac{b_1 s^{n-1} + \ldots + b_n}{s^n + a_1 s^{n-1} + \ldots + a_n} + d.$$

So, for *any* given minimal realization $\{A, B, C, D\}$ there exists an invertible matrix Z_c such that $A_c = Z_c A Z_c^{-1}$, $B_c = Z_c B$, $C_c = C Z_c^{-1}$.

Exercise: Verify that $Z_c = U_c U^{-1}$ where U_c is the controllability matrix, (11.2), of the pair (A_c, B_c).

Transfer function of the plant, $P(s)$, is obtained from the following identities: the denominator polynomial is

$$D_P(s) = \det(sI - A) = \det(sI - A_c) = s^n + a_1 s^{n-1} + \cdots + a_n,$$

and when $d = 0$, the numerator polynomial is

$$N_P(s) = C\mathrm{adj}(sI - A)B = b_1 s^{n-1} + \ldots + b_n,$$

where $\mathrm{adj}(sI - A) = (sI - A)^{-1}\det(sI - A)$, which is an $n \times n$ matrix whose entries are polynomials of degree less than or equal to $(n - 1)$.

11.2 State Feedback

Now consider the special case $y_o(t) = x(t)$ (i.e., the C matrix is identity and all internal state variables are available to the controller) and assume $r(t) = n(t) \equiv 0$. In this section, we study constant controllers of

the form $C(s) = K = [k_n, \ldots, k_1]$. The controller has n inputs, $-x(t)$ which is $n \times 1$ vector, and it has one output; therefore K is a $1 \times n$ vector. The plant input is

$$u(t) = v(t) - Kx(t)$$

and hence, the feedback system is represented by

$$\dot{x}(t) = Ax(t) + B(v(t) - Kx(t)) = (A - BK)x(t) + Bv(t).$$

We define $A_K := A - BK$ as the "A-matrix" of the closed-loop system. Since (A, B) is controllable, it can be shown that the state space system (A_K, B) is controllable too. So, the feedback system poles are the roots of the *closed-loop characteristic polynomial* $\chi_c(s) = \det(sI - (A - BK))$. In the next section, we consider the problem of finding K from a given desired characteristic polynomial $\chi_c(s)$.

11.2.1 Pole Placement

For the special case $A = A_c$ and $B = B_c$, it turns out that $(A_c - B_c K)$ has the same canonical structure as A_c, and hence,

$$\chi_c(s) = \det(sI - (A_c - B_c K)) = s^n + (a_1 + k_1)s^{n-1} + \cdots + (a_n + k_n).$$

Let the desired characteristic polynomial be

$$\chi_c(s) = s^n + \alpha_1 s^{n-1} + \cdots + \alpha_n .$$

Then the controller gains should be set to $k_i = (\alpha_i - a_i)$, $i = 1, \ldots, n$.

Example 11.1 Let the poles of the plant be -1, 0, $2 \pm j1$, then $D_P(s) = s^4 - 3s^3 + s^2 + 5s$. Suppose we want to place the closed-loop system poles to -1, -2, $-2 \pm j1$, i.e., desired characteristic polynomial is $\chi_c(s) = s^4 + 7s^3 + 19s^2 + 23s + 10$. If the system is in controllable

canonical form and we have access to all the states, then the controller $C(s) = [k_4, \ k_3, k_2, k_1]$ solves this pole placement problem with $k_1 = 10$, $k_2 = 18$, $k_3 = 18$ and $k_4 = 10$.

If the system (A, B) is not in the controllable canonical form, then we apply the following procedure:

Step 0. Given A and B, construct the controllability matrix U via (11.2). Check that it is invertible, otherwise closed-loop system poles cannot be placed arbitrarily.

Step 1. Given A, set $D_P(s) = \det(sI - A)$ and calculate the coefficients a_1, \ldots, a_n of this polynomial. Then, define the controllable canonical equivalent of (A, B):

$$A_c = \begin{bmatrix} 0_{(n-1)\times 1} & I_{(n-1)\times(n-1)} \\ -a_n & \cdots & -a_1 \end{bmatrix} \quad B_c = \begin{bmatrix} 0_{(n-1)\times 1} \\ 1 \end{bmatrix}.$$

Step 2. Given desired closed-loop poles, r_1, \ldots, r_n, set
$$\chi_c(s) = (s - r_1) \cdots (s - r_n) = s^n + \alpha_1 s^{n-1} + \cdots + \alpha_n.$$

Step 3. Let $\tilde{k}_i = \alpha_i - a_i$ for $i = 1, \ldots, n$ and $\tilde{K} = [\tilde{k}_n, \ldots, \tilde{k}_1]$. Then, the following controller solves the pole placement problem:

$$C(s) = K = \tilde{K} U_c U^{-1}$$

where U_c is the controllability matrix associated with (A_c, B_c).

Step 4. Verify that the roots of $\chi_c(s) = \det(sI - (A - BK))$ are precisely r_1, \ldots, r_n.

Example 11.2 Consider a system whose A and B matrices are

$$A = \begin{bmatrix} 1 & 0 & 1 \\ 0 & -2 & 2 \\ 0 & 1 & 0 \end{bmatrix} \quad B = \begin{bmatrix} 1 \\ 1 \\ 0 \end{bmatrix}.$$

First we compute $U = [B \vdots AB \vdots A^2 B]$ and check that it is invertible:

$$U = \begin{bmatrix} 1 & 1 & 2 \\ 1 & -2 & 6 \\ 0 & 1 & -2 \end{bmatrix}, \qquad U^{-1} = \begin{bmatrix} 1 & 2 & 5 \\ 1 & -1 & -2 \\ 0.5 & -0.5 & -1.5 \end{bmatrix}.$$

Next, we find $D_P(s) = \det(sI - A) = s^3 + s^2 - 4s + 2$. The roots of $D_P(s)$ are -2.73, 0.73, and 1.0, so the plant is unstable. From the coefficients of $D_P(s)$, we determine A_c and compute U_c:

$$A_c = \begin{bmatrix} 0 & 1 & 0 \\ 0 & 0 & 1 \\ -2 & 4 & -1 \end{bmatrix}, \quad B_c = \begin{bmatrix} 0 \\ 0 \\ 1 \end{bmatrix}, \quad U_c = \begin{bmatrix} 0 & 0 & 1 \\ 0 & 1 & -1 \\ 1 & -1 & 5 \end{bmatrix}.$$

Suppose we want the closed-loop system poles to be at -2, $-1 \pm j1$, then $\chi_c(s) = s^3 + 4s^2 + 6s + 4$. Comparing the coefficients of $\chi_c(s)$ and $D_P(s)$ we find $\tilde{K} = [2 \ 10 \ 3]$. Finally, we compute

$$K = \tilde{K} U_c U^{-1} = [7.5 \quad -4.5 \quad -9.5].$$

Verify that the roots of $\det(sI - (A - BK))$ are indeed -2, $-1 \pm j1$.

When the system is controllable, we can choose the roots of $\chi_c(s)$ arbitrarily. Usually, we want them to be far from the Im-axis in the left half plane. On the other hand, if the magnitudes of these roots are very large, then the entries of K may be very large, which means that the control input $u(t) = -Kx(t)$ (assume $v(t) = 0$ for the sake of argument) may have a large magnitude. To avoid this undesirable situation, we can use linear quadratic regulators.

11.2.2 Linear Quadratic Regulators

In the above setup, assume $v(t) = 0$ and $x(0) = x_o \neq 0$. By a state feedback $u(t) = -Kx(t)$ we want to regulate a fictitious output $\tilde{y}(t) = \tilde{Q}x(t)$

to zero (the $1 \times n$ vector \widetilde{Q} assigns different weights to each component of $x(t)$ and it is assumed that the pair $(A^{\mathsf{T}}, \widetilde{Q}^{\mathsf{T}})$ is controllable), but we do not want to use too much input energy. The trade-off can be characterized by the following cost function to be minimized:

$$J_{\mathrm{LQR}}(K) := \int_0^\infty \left(|\widetilde{y}(t)|^2 + |\widetilde{r}\, u(t)|^2 \right) dt = \|\widetilde{y}\|_2^2 + \|\widetilde{r}\, u\|_2^2 \quad (11.5)$$

where $\widetilde{r} > 0$ determines the relative weight of the control input: when \widetilde{r} is large the cost of input energy is high. The optimal K minimizing the cost function J_{LQR} is

$$\boxed{K_2 := \widetilde{r}^{-2} B^{\mathsf{T}} X} \qquad\qquad (11.6)$$

where X is an $n \times n$ symmetric matrix satisfying the matrix equation

$$A^{\mathsf{T}} X + X A - \widetilde{r}^{-2} X B B^{\mathsf{T}} X + \widetilde{Q}^{\mathsf{T}} \widetilde{Q} = 0 \, ,$$

which can be solved by using the lqr command of MATLAB. The subscript $_2$ is used in the optimal K because J_{LQR} is defined from the $\mathcal{L}_2[0, \infty)$ norms of \widetilde{y} and $\widetilde{r}u$, (11.5).

In this case, the plant is $P(s) = (sI - A)^{-1}B$ and the controller is $C(s) = K$, so the open-loop transfer function is $G(s) = C(s)P(s) = K(sI - A)^{-1}B$. For $K = K_2$, it has been shown that (see e.g., [54]) $G(j\omega)$ satisfies $|1 + G(j\omega)| \geq 1$, for all ω, which means that VM ≥ 1, PM $\geq 60°$, $\mathrm{GM}_{\mathrm{upp}} = \infty$ and $\mathrm{GM}_{\mathrm{low}} \leq \frac{1}{2}$.

Example 11.3 Consider A and B given in Example 11.2 and let $\widetilde{Q} = [1 \quad 4 \quad -2]$, $\widetilde{r} = 1$. Then, by using the lqr command of MATLAB we find $K = [9.3 \quad -4.82 \quad -10.9]$. For $\widetilde{r} = 10$ and $\widetilde{r} = 0.1$, the results are $K = [6.5 \quad -3.0 \quad -6.5]$ and $K = [56.6 \quad -34.6 \quad -84.6]$, respectively. Figure 11.2 shows the closed-loop system poles, i.e. the roots of $\chi_c(s) = \det(sI - (A - BK_2))$, as \widetilde{r}^{-1} increases from 0 to $+\infty$.

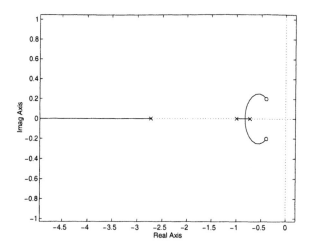

Figure 11.2: Closed-loop poles for an optimal LQR controller.

The figure coincides with the root locus plot of an open-loop system whose poles are $-2.73, -1.0, -0.73$ and zeros are $-0.4 \pm j0.2$. Recall that the roots of $\det(sI - A)$ are $-2.73, 1.0$ and 0.73. Also, check that the roots of $\widetilde{Q}\mathrm{adj}(sI - A)B$ are $0.4 \pm j0.2$. To generalize this observation, let $\tilde{p}_1, \ldots, \tilde{p}_n$ be the poles and $\tilde{z}_1, \ldots, \tilde{z}_m$ be the zeros of $\widetilde{Q}(sI - A)^{-1}B$. Now define $p_i = \tilde{p}_i$ if $\mathrm{Re}(\tilde{p}_i) \leq 0$ and $p_i = -\tilde{p}_i$ if $\mathrm{Re}(\tilde{p}_i) > 0$, for $i = 1, \ldots, n$. Define z_1, \ldots, z_m similarly. Then, the plot of the roots of $\chi_c(s)$, as \tilde{r}^{-1} increases from 0 to $+\infty$, is precisely the root locus of a system whose open-loop poles and zeros are p_1, \ldots, p_n, and z_1, \ldots, z_m, respectively.

Exercise. For the above example, find $G(s) = K(sI - A)^{-1}B$; draw the Nyquist plot, verify the inequality $|1 + G(j\omega)| \geq 1$ and compute the stability margins VM, PM, $\mathrm{GM}_{\mathrm{upp}}$, $\mathrm{GM}_{\mathrm{low}}$ for $\tilde{r} = 0.1, 1.0, 10$.

Another interpretation of the LQR problem is the following. Define $\Psi_1(s) := \widetilde{Q}(sI - (A - BK))^{-1}$, $\Psi_2(s) := \tilde{r}K(sI - (A - BK))^{-1}$ and

$$\widetilde{J}_2(K) := \int_{-\infty}^{\infty} (\|\Psi_1(j\omega)\|^2 + \|\Psi_2(j\omega)\|^2)\mathrm{d}\omega \qquad (11.7)$$

where we have used the notation $\|V\| := \sqrt{|v_1|^2 + \ldots + |v_n|^2}$ for any $1 \times n$ complex vector $V = [v_1, \ldots, v_n]$. Then, the problem of minimizing $J_{LQR}(K)$ is equivalent to minimizing $\tilde{J}_2(K)$ over all state feedback gains K resulting in a stable $\chi_c(s) = \det(sI - (A - BK))$.

A slightly modified optimal state feedback problem is to find K such that the roots of $\chi_c(s)$ are in the open left half plane and

$$\tilde{H}_2 := \int_{-\infty}^{+\infty} (|W_1(j\omega)S(j\omega)|^2 + |W_2(j\omega)T(j\omega)|^2)d\omega \qquad (11.8)$$

is minimized for

$$|W_1(j\omega)| = |\tilde{Q}(j\omega I - A)^{-1}B| \quad \text{and} \quad |W_2(j\omega)| = \tilde{r} \qquad (11.9)$$

where $S(s) = (1+G(s))^{-1}$, $T(s) = 1 - S(s)$ and $G(s) = K(sI - A)^{-1}B$. It turns out that K_2 is also the optimal K minimizing \tilde{H}_2, see [54, pp. 390-391].

11.3 State Observers

In the previous section, we assumed that the state variables are available for feedback. We now return to our original SISO setting, where $y_o(t) = Cx(t)$ is the single output of the plant. The basic idea in state space based control is to generate $\hat{x}(t)$, which is an estimate of $x(t)$ obtained from $y(t)$, and then use $\hat{x}(t)$ in state feedback as if it were the actual $x(t)$.

A state observer is used in generating the state estimate $\hat{x}(t)$:

$$\dot{\hat{x}}(t) = A\hat{x}(t) + Bu(t) + L(y(t) - C\hat{x}(t)) \qquad (11.10)$$

where L is the $n \times 1$ observer gain vector to be designed. The observer (11.10) mimics the plant (11.1) with an additional correction term $L(y(t) - \hat{y}(t))$, where $\hat{y}(t) = C\hat{x}(t)$.

The estimation error $e(t) := x(t) - \widehat{x}(t)$ satisfies

$$\dot{e}(t) = \dot{x}(t) - \dot{\widehat{x}}(t) = (A - LC)e(t).$$

Therefore, $e(t)$ is the inverse Laplace transform of $(sI-(A-LC))^{-1}e(0)$, where $e(0)$ is the initial value of the estimation error. We select L in such a way that the roots of $\chi_o(s) = \det(sI - (A - LC))$ are in the open left half plane. Then, $e(t) \to 0$ as $t \to \infty$. In fact, the rate of decay of $e(t)$ is related to the location of the roots of $\chi_o(s)$. Usually, these roots are chosen to have large negative real parts so that $\widehat{x}(t)$ converges to $x(t)$ very fast. Note that

$$\chi_o(s) = \det(sI - (A - LC)) = \det(sI - (A^{\mathsf{T}} - C^{\mathsf{T}}L^{\mathsf{T}})).$$

Therefore, once the desired roots of $\chi_o(s)$ are given, we compute L as $L = K^{\mathsf{T}}$, where K is the result of the pole placement procedure with the data $\chi_c \leftarrow \chi_o$, $A \leftarrow A^{\mathsf{T}}$, $B \leftarrow C^{\mathsf{T}}$. For arbitrary placement of the roots of $\chi_o(s)$, the pair $(A^{\mathsf{T}}, C^{\mathsf{T}})$ must be controllable.

Exercise. Let $C = \begin{bmatrix} 1 & -4 & 2 \end{bmatrix}$ and consider the A matrix of Example 11.2. Find the appropriate observer gain L, such that the roots of $\chi_o(s)$ are -5, $-8 \pm j2$.

11.4 Feedback Controllers

11.4.1 Observer Plus State Feedback

Now we use state feedback in the form $u(t) = v(t) - K\widehat{x}(t)$, where $\widehat{x}(t)$ is generated by the observer (11.10). The key assumption used in (11.10) is that both $y(t)$ and $u(t)$ are known exactly, i.e., there are no disturbances or measurement noises. But $v(t)$ is a disturbance that enters the observer formula via $u(t)$. Also, $y(t) = y_o(t) + n(t)$, where $y_o(t)$ is the actual plant output and $n(t)$ is measurement noise. We should

not ignore the reference input $r(t)$ either. Accordingly, we modify the observer equations and define the controller as

$$\text{Controller}: \begin{cases} \dot{\widehat{x}}(t) = \widehat{A}\widehat{x}(t) - L(r(t) - y(t)) \\ u(t) = -K\widehat{x}(t) + v(t) \end{cases}$$

where $\widehat{A} := (A - BK - LC)$. The input to the controller is $(r(t) - y(t))$ and the controller output is $-K\widehat{x}(t)$. Transfer function of the feedback controller is

$$C(s) = \frac{N_C(s)}{D_C(s)} = K(sI - \widehat{A})^{-1}L. \tag{11.11}$$

In this case, the state estimation error $e(t) = x(t) - \widehat{x}(t)$ satisfies

$$\dot{e}(t) = (A - LC)e(t) + Bv(t) + L(r(t) - n(t)). \tag{11.12}$$

Moreover, the state $x(t)$ can be determined from $e(t)$:

$$\dot{x}(t) = (A - BK)x(t) + BKe(t) + Bv(t). \tag{11.13}$$

Equations (11.12) and (11.13) determine the feedback system behavior: closed-loop system poles are the roots of $\chi_c(s) = \det(sI - (A - BK))$ and $\chi_o(s) = \det(sI - (A - LC))$. If the state space realization $\{A, B, C\}$ of the plant is minimal, then by proper choices of K and L closed-loop system poles can be placed arbitrarily.

In general, the controller (11.11) has n poles: these are the roots of $D_C(s) = \det(sI - \widehat{A})$. The zeros of the controller are the roots of the numerator polynomial $N_C(s) = K \text{ adj}(sI - \widehat{A}) L$, where $\text{adj}(\cdot)$ denotes the adjoint matrix which appears in the computation of the inverse. Therefore, the controller is of the same order as the plant $P(s)$, unless K and L are chosen in a special way that leads to pole zero cancelations within $C(s) = N_C(s)/D_C(s)$.

Exercise: By using equations (11.12) and (11.13) show that the closed-loop transfer functions are given by

$$
\begin{aligned}
S(s) &= (1 - KM_1(s)B)\,(1 + KM_2(s)B) \\
T(s) &= CM_1(s)B\,KM_2(s)L \\
C(s)S(s) &= (1 - KM_1(s)B)\,KM_2(s)L \\
P(s)S(s) &= CM_1(s)B\,(1 + KM_2(s)B)
\end{aligned}
$$

where $M_1(s) := (sI - (A - BK))^{-1}$, $M_2(s) := (sI - (A - LC))^{-1}$, $S(s) = (1 + P(s)C(s))^{-1}$ and $T(s) = 1 - S(s)$.

11.4.2 \mathcal{H}_2 Optimal Controller

We have seen that by appropriately selecting K and L a stabilizing controller can be constructed. As far as closed-loop system stability is concerned, designs of K and L are decoupled. Let us assume that K is determined from an LQR problem as the optimal gain $K = K_2$ associated with the problem data \widetilde{Q} and \tilde{r}. Now consider the state estimation error $e(t)$ given by (11.12). Suppose $r(t) = 0$, $v(t) \neq 0$ and $n(t) = -\tilde{q}w(t) \neq 0$ for some constant $\tilde{q} > 0$. Then, the transfer function from v to e is $\Phi_1(s) = (sI - (A - LC))^{-1}B$ and the transfer function from w to e is $\Phi_2(s) = \tilde{q}(sI - (A - LC))^{-1}L$. Define

$$
\widehat{J}_2(L) := \int_{-\infty}^{\infty} (\|\Phi_1(j\omega)^{\mathsf{T}}\|^2 + \|\Phi_2(j\omega)^{\mathsf{T}}\|^2)\,\mathrm{d}\omega. \tag{11.14}
$$

Comparing (11.14) with (11.7), we see that the problem of minimizing $\widehat{J}_2(L)$ over all L resulting in a stable $\chi_o(s) = \det(sI - (A - LC))$ is an LQR problem with the modified data $A \leftarrow A^{\mathsf{T}}$, $B \leftarrow C^{\mathsf{T}}$, $\widetilde{Q} \leftarrow B^{\mathsf{T}}$, $\tilde{r} \leftarrow \tilde{q}$, $K \leftarrow L^{\mathsf{T}}$. Hence, the optimal solution is $L = L_2$:

$$
\boxed{L_2 = \tilde{q}^{-2}YC^{\mathsf{T}}} \tag{11.15}
$$

the $n \times n$ symmetric matrix Y satisfies

$$AY + YA^{\mathsf{T}} - \tilde{q}^{-2}YC^{\mathsf{T}}CY + BB^{\mathsf{T}} = 0.$$

We can use the `lqr` command of MATLAB (with the data as specified above) to find Y and corresponding L_2.

The controller $C_{2\mathrm{opt}}(s) = K_2(sI - (A - BK_2 - L_2C))^{-1}L_2$ is an \mathcal{H}_2 optimal controller, see [54, Chapter 14]. As $\tilde{q} \searrow 0$, the optimal \mathcal{H}_2 controller $C_{2\mathrm{opt}}(s)$ approaches the solution of the following problem: find a stabilizing controller $C(s)$ for $P(s) = C(sI - A)^{-1}B$, such that \tilde{H}_2, (11.8), is minimized, where $S(s) = (1 + G(s))^{-1}$, $T(s) = 1 - S(s)$, $G(s) = C(s)P(s)$ and the weights are defined by (11.9). Recall that in (10.8) we try to minimize the peak value of

$$\Psi(\omega) := |W_1(j\omega)S(j\omega)|^2 + |W_2(j\omega)T(j\omega)|^2 \tag{11.16}$$

whereas here we minimize the integral of $\Psi(\omega)$.

Example 11.4 Consider the plant

$$P(s) = \frac{s - 1}{s^3 + s^2 - 4s + 6}$$

with controllable canonical realization

$$A = \begin{bmatrix} 0 & 1 & 0 \\ 0 & 0 & 1 \\ -6 & 4 & -1 \end{bmatrix} \quad B = \begin{bmatrix} 0 \\ 0 \\ 1 \end{bmatrix} \quad C = [-1 \ 1 \ 0] \, .$$

Let $\tilde{Q} = [1 \ -1 \ 1]$, $\tilde{r} = 2$ and $\tilde{q} = 0.1$. Using the `lqr` command of MATLAB we compute the optimal \mathcal{H}_2 controller as outlined above. Verify that for this example the controller is

$$C_{2\mathrm{opt}}(s) = \frac{20.6 \, (s + 3) \, (s + 10)}{(s + 5) \, (s^2 + 5.34 \, s + 22.52)}$$

and the closed-loop system poles are $\{-2.6 \pm j1.7, -3.0, -1 \pm j1, -1.2\}$.

Exercise. Find the poles and zeros of $P(s)$ and $\widetilde{Q}(sI - A)^{-1}B$. Draw the root locus (closed-loop system pole locations) in terms of \tilde{r}^{-1} and \tilde{q}^{-1}. Obtain the Nyquist plot of $G(j\omega) = C_{2\text{opt}}(j\omega)P(j\omega)$ for the above example. Compute the vector margin and compare it with the vector margin of the optimal LQR system.

11.4.3 Parameterization of all Stabilizing Controllers

Given a strictly proper SISO plant $P(s) = C(sI - A)^{-1}B$, the set of all controllers stabilizing the feedback system, denoted by \mathcal{C}, is parameterized as follows. First, find two vectors K and L such that the roots of $\chi_c(s) = \det(sI - (A - BK))$ and $\chi_o(s) = \det(sI - (A - LC))$ are in the open left half plane (the pole placement procedure can be used here). For any proper stable $Q(s)$ introduce the notation

$$C_Q(s) \quad := \quad KM(s)L + \frac{(1 - KM(s)B)(1 - CM(s)L)\, Q(s)}{1 - CM(s)B\, Q(s)} \quad (11.17)$$

$$= \quad \frac{KM_2(s)L + (1 - CM_2(s)L)\, Q(s)}{1 + KM_2(s)B - CM_2(s)B\, Q(s)} \quad (11.18)$$

where $M_2(s) := (sI - (A - LC))^{-1}$, $M(s) := (sI - \widehat{A})^{-1}$ and $\widehat{A} := A - BK - LC$. Then, the set \mathcal{C} is given by

$$\mathcal{C} = \{\, C_Q(s) \quad : \quad Q \text{ is proper and stable}\} \quad (11.19)$$

where $C_Q(s)$ is defined in (11.17) or (11.18). For $Q(s) = 0$, we obtain $C(s) = C_0(s) = K(sI - \widehat{A})^{-1}L$, which coincides with the controller expression (11.11). The parameterization (11.19) is valid for both stable and unstable plants. In the special case where the plant is stable, we can choose $K = L^{\mathsf{T}} = 0_{1 \times n}$, which leads to $\widehat{A} = A$ and $CM(s)B(s) = P(s)$, and hence, (11.19) becomes the same as (10.11). A block diagram of the feedback system with controller in the form (11.17) is shown in

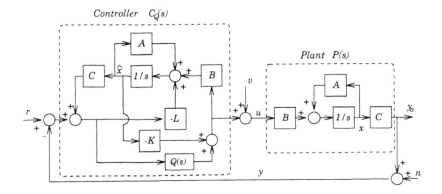

Figure 11.3: Feedback system with a stabilizing controller.

Figure 11.3. The parameterization (11.19) is a slightly modified form of the Youla parameterization, [54]. An extension of this parameterization for nonlinear systems is given in [35].

11.5 Exercise Problems

1. Consider the pair

$$A = \begin{bmatrix} 0 & 1 & 0 \\ 1 & -2 & 1 \\ 0 & 0 & -3 \end{bmatrix} \quad B = \begin{bmatrix} 0 \\ 1 \\ 0 \end{bmatrix}.$$

What are the roots of $D_P(s) := \det(sI - A) = 0$? Show that (A, B) is not controllable. However, there exists a 1×3 vector K such that the roots of $\chi_c(s) = \det(sI - (A - BK))$ are $\{r_1, r_2, -3\}$ where r_1 and r_2 can be assigned arbitrarily. Let $r_{1,2} = -2 \pm j1$ and find an appropriate gain K. In general, the number of assignable poles is equal to the rank of the controllability matrix. Verify that the controllability matrix has rank two in this example. If the non-assignable poles are already in the left half plane, then the pair (A, B) is said to be *stabilizable*. In this example, the pole at

−3 is the only non-assignable pole, so the system is stabilizable.

2. Verify that the transfer function of the controller shown in Figure 11.3 is equal to $C_Q(s)$ given by (11.17) and (11.18).

3. Show that the closed-loop transfer functions corresponding to the feedback system of Figure 11.3 are:

$$
\begin{aligned}
S(s) &= (1 - KM_1(s)B)\,(1 + KM_2(s)B - CM_2(s)B\,Q(s)) \\
T(s) &= CM_1(s)B\,(KM_2(s)L + (1 - CM_2(s)L)\,Q(s)) \\
C(s)S(s) &= (1 - KM_1(s)B)\,(KM_2(s)L + (1 - CM_2(s)L)\,Q(s)) \\
P(s)S(s) &= CM_1(s)B\,(1 + KM_2(s)B - CM_2(s)B\,Q(s))
\end{aligned}
$$

where $M_1(s) = (sI-(A-BK))^{-1}$ and $M_2(s) = (sI-(A-LC))^{-1}$.

4. Given

$$
P(s) = \frac{(s - z)}{(s + z)(s - 1)} \qquad z > 0 \ \text{ and } \ z \neq 1
$$

obtain the controllable canonical realization of $P(s)$.
Let $\tilde{Q} = [10z \ \ 1]$, $\tilde{r} = 1$ and $\tilde{q} = 0.1$.

(i) Compute the \mathcal{H}_2 optimal controllers for $z = 0.1$, 0.9, 1.1, 15. Determine the poles and zeros of these controllers and obtain the Nyquist plots. What happens to the vector margin as $z \to 1$?

(ii) Plot the corresponding $\Psi(\omega)$ for each of these controllers. What happens to the peak value of $\Psi(\omega)$ as $z \to 1$?

(iii) Fix $z = 4$ and let $K = K_2$, $L = L_2$. Determine the closed-loop transfer functions S, T, CS and PS whose general structures are given in Problem 3. Note that

$$
P(s) = \frac{P(s)S(s)}{S(s)} = \frac{T(s)}{C(s)S(s)} = \frac{CM_1(s)B}{1 - KM_1(s)B}.
$$

Verify this identity for the \mathcal{H}_2 control problem considered here.

5. Difficulty of the controller design is manifested in parts (i) and (ii) of the above exercise problem. We see that if the plant has a right half plane pole and a zero close by, then the vector margin is very small. Recall that

$$\text{VM}^{-1} = \|S\|_\infty$$

and $S(s) = (1 - KM_1(s)B)(1 + KM_2(s)B - CM_2(s)B\,Q(s))$.

(i) Check that, for the above example, when $z = 0.9$ we have $S(s) = \frac{(s-1)}{(s+1)}S_0(s)$ where

$$S_0(s) = \frac{(s+1)(s+0.9)}{s^2 + 4.57\,s + 9}\;\frac{(s+47.88)(s-32.26) - (s-0.9)Q(s)}{s^2 + 10.95\,s + 9.05}$$

Clearly $\|S\|_\infty = \|S_0\|_\infty$. Thus,

$$\text{VM}^{-1} = \sup_{\text{Re}(s)>0} |S_0(s)| \geq |S_0(0.9)| \cong 19 \;,$$

which means that $\text{VM} \leq 0.0526$ for all proper stable $Q(s)$, i.e., the vector margin will be less than 0.0526 no matter what the stabilizing controller is. This is a limitation due to plant structure.

(ii) Find a proper stable $Q(s)$ which results in $\text{VM} \geq 0.05$. Hint: write $S_0(s)$ in the form

$$S_0(s) = S_1(s) - S_2(s)Q(s)$$

where $S_1(s)$ and $S_2(s)$ are proper and stable and $S_2(s)$ has only one zero in the right half plane at $z = 0.9$. Then set

$$Q(s) = Q_0(s) := \frac{1}{(1+\epsilon s)^\ell}\;\frac{S_1(s) - S_1(0.9)}{S_2(s)}$$

where ℓ is relative degree of $S_2(s)$ and $\epsilon \searrow 0$. Check that $Q(s)$ defined this way is proper and stable, and it leads to $\|S_0\|_\infty \leq 20$ for sufficiently small ϵ. Finally, the controller is obtained by just plugging in $Q(s)$ in (11.17) or (11.18).

Bibliography

[1] Akçay, H., G. Gu, and P. P. Khargonekar, "Class of algorithms for identification in H_∞: continuous-time case," *IEEE Transactions on Automatic Control*, vol. 38 (1993), pp. 289-296.

[2] Anantharam, V., and C. A. Desoer, "On the stabilization of nonlinear systems," *IEEE Trans. on Automatic Control*, vol. 29 (1984), pp. 569-572.

[3] Antsaklis, P. J., and A. N. Michel, *Linear Systems*, McGraw-Hill, New York, 1997.

[4] Balas, G. J., J. C. Doyle, K. Glover, A. Packard, and R. Smith, *μ-Analysis and Synthesis Toolbox User's Guide*, MathWorks Inc., Natick MA, 1991.

[5] Barmish, B. R., *New Tools for Robustness of Linear Systems*, Macmillan, New York, 1994.

[6] Bay, J. S., *Fundamentals of Linear State Space Systems*, McGraw-Hill, New York, 1999.

[7] Bélanger, P. R., *Control Engineering*, Saunders College Publishing, Orlando, 1995.

[8] Bellman, R. E., and K. L. Cooke, *Differential Difference Equations*, Academic Press, New York, 1963.

[9] Bhattacharyya, S. P., H. Chapellat, and L. H. Keel, *Robust Control, The Parametric Approach*, Prentice-Hall, Upper Saddle River NJ, 1995.

[10] Bode, H. W., *Network Analysis and Feedback Amplifier Design*, D. Van Nostrand Co., Inc., Princeton NJ, 1945.

[11] Chen, C. T., *Linear System Theory and Design*, 3rd ed., Oxford University Press, New York, 1999.

[12] Curtain, R. F., "A synthesis of time and frequency domain methods for the control of infinite-dimensional systems: a system theoretic approach," *Control and Estimation in Distributed Parameter Systems*, H. T. Banks ed., SIAM, Philadelphia, 1992, pp. 171–224.

[13] Curtain, R. F., and H. J. Zwart, *An Introduction to Infinite-Dimensional Linear System Theory*, Springer-Verlag, New York, 1995.

[14] Davison, E. J., and A. Goldenberg, "Robust control of a general servomechanism problem: the servo compensator," *Automatica*, vol. 11 (1975), pp. 461–471.

[15] Desoer, C. A., and M. Vidyasagar *Feedback Systems: Input-Output Properties*, Academic Press, New York, 1975.

[16] Devilbiss, S. L., and S. Yurkovich, "Exploiting ellipsoidal parameter set estimates in H_∞ robust control design," *Int. J. Control*, vol. 69 (1998), pp. 271–284.

[17] Dorf, R. C., and R. H. Bishop, *Modern Control Systems*, 8th ed., Addison Wesley Longman, Menlo Park, 1998.

[18] Doyle, J. C., B. A. Francis, and A. Tannenbaum, *Feedback Control Theory*, Macmillan, New York, 1992.

[19] Etter, D. M., *Engineering Problem Solving with MATLAB*, 2nd ed., Prentice-Hall, Upper Saddle River NJ, 1997.

[20] Foias, C., H. Özbay, and A. Tannenbaum, *Robust Control of Infinite Dimensional Systems*, LNCIS 209, Springer-Verlag, London, 1996.

[21] Francis, B., O. A. Sebakhy and W. M. Wonham, "Synthesis of multivariable regulators: the internal model principle," *Applied Mathematics & Optimization*, vol. 1 (1974), pp. 64–86.

[22] Franklin, G. F., J. D. Powell, and A. Emami-Naeini, *Feedback Control of Dynamic Systems*, 3rd ed., Addison-Wesley, Reading MA, 1994.

[23] Hanselman, D. , and B. Littlefield, *Mastering* MATLAB *5*, Prentice-Hall, Upper Saddle River NJ, 1998.

[24] Hara, S., Y. Yamamoto, T. Omata, and M. Nakano, "Repetitive control system: a new type servo system for periodic exogenous signals," *IEEE Transactions on Automatic Control*, vol. 33 (1988) pp. 659–667.

[25] Helmicki, A. J., C. A. Jacobson and C. N. Nett, "Worst-case/deterministic identification in H_∞: The continuous-time case," *IEEE Transactions on Automatic Control*, vol. 37 (1992), pp. 604–610.

[26] Hemami, H., and B. Wyman, "Modeling and control of constrained dynamic systems with application to biped locomotion in the frontal plane," *IEEE Trans. on Automatic Control*, vol. 24 (1979), pp. 526-535.

[27] Isidori, A., *Nonlinear Control Systems: An Introduction*, 2nd ed., Springer-Verlag, Berlin, 1989.

[28] Kamen, E. W., and B. S. Heck, *Fundamentals of Signals and Systems*, Prentice-Hall, Upper Saddle River NJ, 1997.

[29] Kataria, A., H. Özbay, and H. Hemami, "Point to point motion of skeletal systems with multiple transmission delays," *Proc. of the 1999 IEEE International Conference on Robotics and Automation*, Detroit MI, May 1999, pp. 53–58.

[30] Khajepour, A., M. F. Golnaraghi, and K. A. Morris, "Application of center manifold theory to regulation of a flexible beam," *Journal of Vibration and Acoustics, Trans. of the ASME*, vol. 119 (1997), pp. 158–165.

[31] Khalil, H., *Nonlinear Systems*, 2nd ed., Prentice-Hall, Upper Saddle River NJ, 1996.

[32] Kosut, R. L., M. K. Lau, and S. P. Boyd, "Set-membership identification of systems with parametric and nonparametric uncertainty," *IEEE Transactions on Automatic Control*, vol. 37 (1992), pp. 929–941.

[33] Lam, J., "Convergence of a class of Padé approximations for delay systems," *Int. J. Control*, vol. 52 (1990), pp. 989–1008.

[34] Lenz, K. and H. Özbay, "Analysis and robust control techniques for an ideal flexible beam," in *Multidisciplinary Engineering Systems: Design and Optimization Techniques and their Applications*, C. T. Leondes ed., Academic Press Inc., 1993, pp. 369–421.

[35] Lu, W-M. "A state-space approach to parameterization of stabilizing controllers for nonlinear systems," *IEEE Trans. on Automatic Control*, vol. 40 (1995), pp. 1579-1588.

[36] Mäkilä, P. M., J. R. Partington, and T. K. Gustafsson, "Worst-case control-relevant identification," *Automatica*, vol. 31 (1995), pp. 1799–1819

[37] Ogata, K., *Modern Control Engineering*, 3rd ed., Prentice-Hall, Upper Saddle River NJ, 1997.

[38] Oppenheim, A. V. and A. S. Willsky, with S. H. Nawab, *Signals and Systems*, 2nd ed., Prentice-Hall, Upper Saddle River NJ, 1997.

[39] Özbay, H., "Active control of a thin airfoil: flutter suppression and gust alleviation," *Preprints of 12th IFAC World Congress*, Sydney Australia, July 1993, vol. 2, pp. 155–158.

[40] Özbay, H., and G. R. Bachmann, "H^2/H^∞ Controller design for a 2D thin airfoil flutter suppression," *AIAA Journal of Guidance Control & Dynamics*, vol. 17 (1994), pp. 722–728.

[41] Özbay, H., S. Kalyanaraman, A. İftar, "On rate-based congestion control in high-speed networks: Design of an H^∞ based flow controller for single bottleneck," *Proceedings of the American Control Conference*, Philadelphia PA, June 1998, pp. 2376–2380.

[42] Özbay, H., and J. Turi, "On input/output stabilization of singular integro-differential systems," *Applied Mathematics and Optimization*, vol. 30 (1994), pp. 21–49.

[43] Peery, T. E. and H. Özbay, "H^∞ optimal repetitive controller design for stable plants," *Transactions of the ASME Journal of Dynamic Systems, Measurement, and Control*, vol. 119 (1997), pp. 541–547.

[44] Pratap, R., *Getting Started with MATLAB 5*, Oxford University Press, Oxford 1999.

[45] Rohrs C. E., J. L. Melsa, and D. G. Schultz, *Linear Control Systems*, McGraw-Hill, New York, 1993.

[46] Russell, D. L., "On mathematical models for the elastic beam with frequency-proportional damping," in *Control and Estimation in Distributed Parameter Systems*, H. T. Banks ed., SIAM, Philadelphia, 1992, pp. 125–169.

[47] Rugh, W. J. *Linear System Theory*, 2nd ed., Prentice-Hall, Upper Saddle River NJ, 1996.

[48] Smith, O. J. M., *Feedback Control Systems*, McGraw-Hill, New York, 1958.

[49] Stepan, G., *Retarded Dynamical Systems: Stability and Characteristic Functions*, Longman Scientific & Technical, New York, 1989.

[50] Toker, O. and H. Özbay, "H^∞ optimal and suboptimal controllers for infinite dimensional SISO plants," *IEEE Transactions on Automatic Control*, vol. 40 (1995), pp. 751–755.

[51] Ulus, C., "Numerical computation of inner-outer factors for a class of retarded delay systems," *Int. Journal of Systems Sci.*, vol. 28 (1997), pp. 897–904.

[52] Van Loan, C. F., *Introduction to Scientific Computing: A Matrix-Vector Approach Using* MATLAB, Prentice-Hall, Upper Saddle River NJ, 1997.

[53] Youla, D. C., H. A. Jabr, and J. J. Bongiorno Jr., "Modern Wiener Hopf design of optimal controllers: part II," *IEEE Transactions on Automatic Control*, vol. 21 (1976), pp. 319–338.

[54] Zhou, K., with J. C. Doyle, and K. Glover, *Robust and Optimal Control*, Prentice-Hall, Upper Saddle River NJ, 1996.

Index

215

Milton Keynes UK
Ingram Content Group UK Ltd.
UKHW040101071024
449327UK00019B/728